THE SCIENCE OF
UFOs

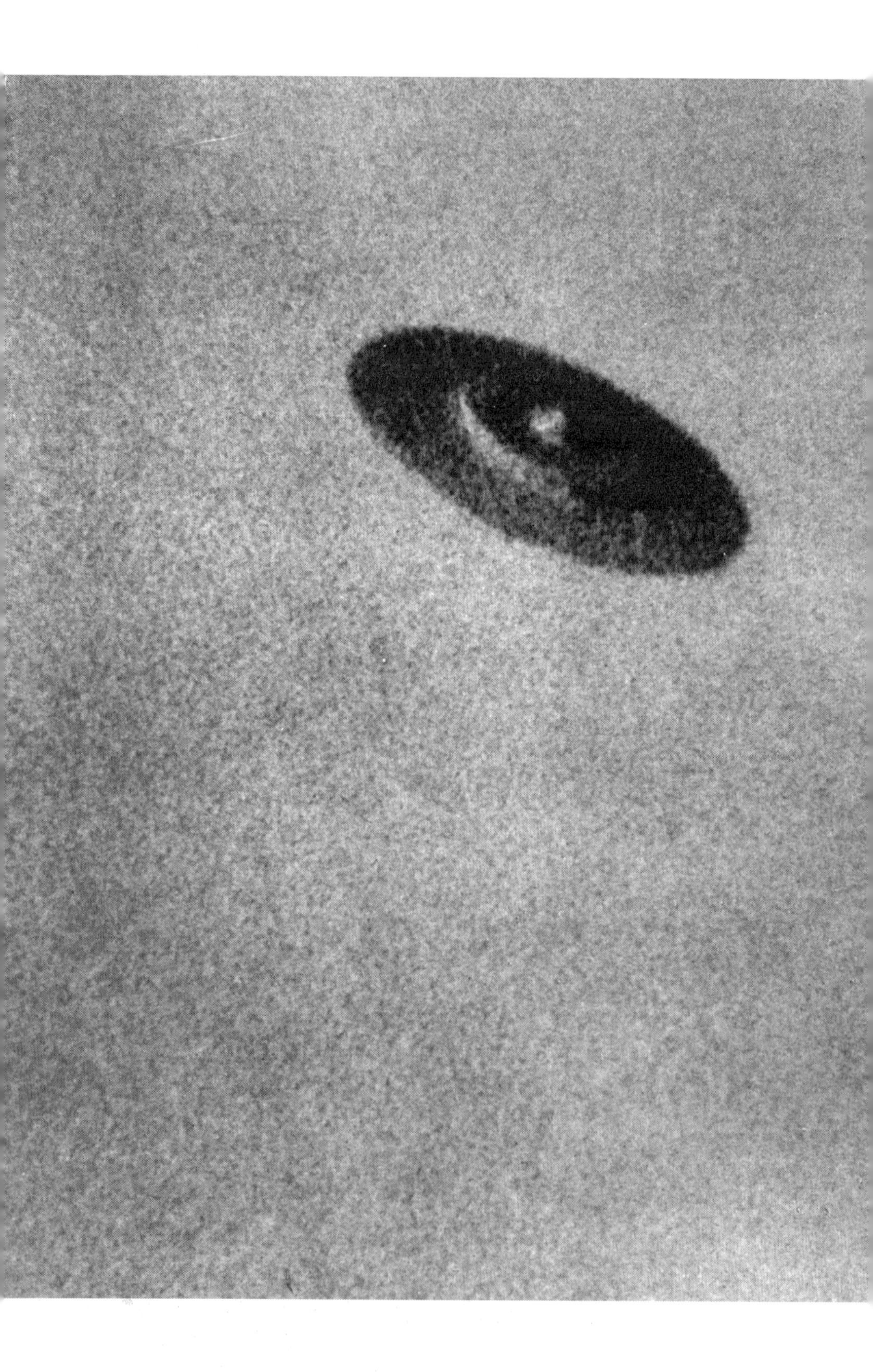

WILLIAM R. ALSCHULER, PH.D.

Edited by Howard Zimmerman

A Byron Preiss Book

St. Martin's Press
New York

The author dedicates this book to his daughter,

Rachel B. Alschuler.

Keep your camera dry, your film cool,

and your mind open.

A Byron Preiss Book.

www.stmartins.com

Library of Congress Cataloging-in-Publication Data

Alschuler, William R.
The science of UFOs / William R. Alschuler ; edited by Howard Zimmerman.—1st ed.
p. cm.
Includes bibliographical references.
ISBN 0-312-26225-6
1. Unidentified flying objects—Sightings and encounters. 2. Explanation.
3. Science—Theory reduction. I. Zimmerman, Howard. II. Title.

TL789 .A562 2001
001.942—dc21 00-045760

Cover design by j.vita
Interior design by Michael Mendelsohn/MM Design 2000

First Edition: February 2001

10 9 8 7 6 5 4 3 2 1

TABLE OF CONTENTS

ACKNOWLEDGMENTS

The author would like to offer thanks to many people. For general support: Dean Dick Hebdige, School of Critical Studies, California Institute of the Arts; for rapid responses to specific special requests, his assistants, Cathi Love and Lisa Petersen. For her humor and patience in the face of continued postponements and schedule rearrangements and occasional snappishness by the author: his daughter Rachel. For long-distance support, his daughter Elise. For continuous support of every sort, patience, creativity, love and understanding: his wife, Karen. To Byron Preiss for the chance to do this book.

To my editor, Howard Zimmerman, the *deus ex machina* who provided guidance, extraordinary expertise, comeuppance and humor at every critical moment: extreme thanks . . . and last writes.

PREFACE

Say the word "UFOs" or the phrase "alien abductions" in some circles, and there is a knowing snicker. In others, there is an emotionally charged silence. These days the words carry heavy baggage in western industrial societies, especially in the United States. There are now at least three industries that have grown up around them: an entertainment industry in the form of science-fiction books and movies; a second entertainment industry in the form of made-for-TV movies that purport to examine the reality of the associated phenomena; and a small industry that consists of support groups and therapists who counsel those who say aliens have abducted them, in some cases for non-consensual experiments. This goes to show that, among other things, in America one can commercialize almost anything, including what is often delusion or lies, as well as the pain of innocent victims (self-inflicted or not).

If you think the above statement means that I have a closed mind toward the phenomena of UFOs and alien abductions, you are wrong. I am just skeptical of them. I have been enthusiastically interested in UFOs since the time in childhood when I began to consume science fiction and science fact in great gulps. The number of sci-fi stories I have read on the subject is quite large, and I went on to train in astronomy up through getting a doctorate, in part inspired by all that science fiction. I believe it is highly likely that extraterrestrial beings of great intelligence populate our galaxy on many planets and talk to one another across the immense distances between the stars. I am extremely doubtful that they visit one another or that any have visited Earth. But if aliens arrived on my doorstep I would be thrilled, and assuming they didn't flash weapons, I would try to touch hands or any other appropriate organ or limb.

It would of course be big news, the biggest of the new millennium, if it could be shown irrefutably that aliens have visited us. That people from many walks of life have described UFO encounters or abductions and that these descriptions often have common elements is cited as evidence by believers as proof that something real is going on. For centuries, people who sought the origin of meteors felt that rocks could not fall from the sky. Nineteenth-century scientists said the platypus could not exist. For decades, "well-informed scientists" thought it would be impossible to fly faster than the speed of sound. In each case confirming or damning evidence was produced in the form of the objects in question or the performance of the feat. Specimens of the platypus could be collected dead or alive, and live ones could be maintained in a zoo, where they could be viewed on demand and examined in detail. The sonic booms of the supersonic shock wave-riders prove they are traveling faster than sound. UFOs and abductions have remained irreproducible phenomena, without supporting, accepted physical evidence. This sort of situation is one that science—which exists in part precisely for the evaluation of claims about the physical world—handles poorly. It gives scientists heartburn, headaches, and nausea! The reaction of most is to say to believers, "Take your tales elsewhere. Don't waste my time!"

There have been exceptions. Dr. Allen Hynek, professor of astronomy at Northwestern University, whom I knew when I was in high school, took the time to examine systematically a government compilation of the first fifteen years of UFO reports. (I will touch on his results later in the book.) Carl Sagan, professor of astronomy at Harvard and Cornell, whom I knew in college, in his wonderful and tireless efforts to promote rational discourse and a clear view of humanity's position in the cosmos was willing to discuss believers' claims almost up to the day he died in 1998.

I am interested to follow modestly in Hynek's footsteps and try to answer this question: If we organize all the published modern accounts and related matters from older times, what do we have in hand? And having done that work, what does modern science have to say about the material's implications in terms of today's physics, biology, and psychology, and careful extrapolations of them into the future?

This sort of task has been undertaken by others. For example, Professor

The late Dr. J. Allen Hynek in 1987. He lent an air of legitimacy to the study of UFO phenomena. Photo: copyright 2000 John P. Timmerman/CUFOS

Larry Krauss's book, *The Physics of Star Trek*, looks at the physics of that science-fictional universe and comments on what could be and what can't, according to known and speculative physics. Professor Michio Kaku's *Hyperspace* examines the attempt to find a theory of everything in terms of the mathematics of spaces with more than Einstein's four dimensions (three space and one time). Both of these books cover territory relevant here and cover it well. But my landscape overlaps and extends beyond their borders.

I teach a related course, always popular at my college, on the search for extraterrestrial intelligence, in which it is natural to discuss astronomy, physics, evolution, planetary sciences, signal detection strategies and technologies—all of which I love. And I have enjoyed the investigations I describe here to try to make sense of UFO-land. I hope this book will bring some clarity to the subject and a feeling of balanced understanding to readers.

Of course, if tomorrow at dawn an alien ship lands on my street, I will happily report to you the details and eat any words printed here that are proved to be wrong.

William R. Alschuler
San Francisco, 2000

1

EXAMINING THE EVIDENCE

In early December 1998 thousands of people in Jalisco and Aguascaliente States, about 275 miles west of Mexico City, saw "a very bright white light" fly over the Sierra Madre mountains at about 7P.M.

On November 18, 1998 a man in Crosslanes, West Virginia was going into his garage when he happened to look up and spotted a 727 flying overhead: "It was high up enough it left a very clear contrail. Suddenly, at very close range above the plane, a silver-gray disc appeared. It flew directly over the top of the plane and was making a side-to-side swaying motion, almost like a leaf falling. As it moved the Sun glinted off it, and it shone very brightly. Suddenly from the east, a jet (delta shape) came screaming toward the UFO. At that point it had stopped, and the passenger plane had already moved away.

"The UFO glowed very brightly and then shot straight up and was out of sight in seconds. I have an Air Force Intelligence background and what I saw is nothing I am familiar with."

(Reported in *UFO Magazine,* March/April, 1999.)

In Zanesville, Ohio, a small town not far from the infamous Wright-Patterson Air Force Base of Roswell crash fame, on November 13, 1966, Ralph Ditter took two photographs of a UFO. They show a suburban scene with a house and driveway, several parked cars, a bare tree in the foreground and woods behind. It seems to have been a bright day as there are strong shadows. In each photo there is a mechanical looking object hovering over the landscape. Its shape is like that of a top hat whose top has been cut off about a third of the way up. It glints in the sunlight. It casts no shadow. The craft matches descriptions of a number of 1998 sightings made around Zanesville, of which at least two were made by law enforcement officers. The photos have no negatives as they were made with a Polaroid camera. The ships are seen in silhouette against a bright cloudless sky so there is no easy way to estimate their size or distance. They remain unexplained to this day.

(Reported in *Popular Mechanics,* July 1998)

COSMIC PRECIOUS GEMS SCATTERED ON THE ROAD: WHAT DOES IT MEAN TO HAVE EVIDENCE OF UNUSUAL EVENTS?

Many of the people whose memories of being kidnapped by aliens are recounted in the book *Abduction,* by Harvard psychologist John Mack, report being taken aboard alien ships and examined with metal instruments in rooms lit from all surfaces. They also recall aliens of varied descriptions, as if of distinct species. Certain details of these accounts are remarkably similar to each other, and the aliens' reported behavior makes these experiences sometimes shockingly invasive of the abductees' personal space, physical bodies, and sense of sanity.

The above examples of extraordinary events fall into three major categories of what might be called "alien contact." The first are accounts of lights or colors or indistinct shapes in the sky that travel at unbelievable speeds, accelerate at unbelievable rates, appear and disappear while standing still, and generally show behavior not achievable by human aircraft or rocket craft. In some cases their performance seems to violate known laws of physics. These sightings often take place at night, and most of the objects are reported to be self-luminous. These phenomena are called UFOs (Unidentified Flying

Objects) precisely because their nature is unknown until and unless someone comes along to identify or explain them. The Mexico and West Virginia sightings mentioned above fall into this category.

The Zanesville sightings and photos are in a second category: the observers see, describe, and sometimes make photos or videos of a definite shape, one that could be classified as a spacecraft or at least a manufactured object. These, too, behave in ways that human technology cannot achieve, and they are most often daytime sightings. (Of course there are exceptions. In some cases lighting onboard or on the ground allows observers to see details of the craft even at night.)

The stories of interactions with aliens, the details of the insides of their spacecraft, and the tools for physical exams of abductees represent a third level of involvement of the witnesses. Often these events take place at night, perhaps because the abduction experiences frequently start while abductees are in a dream state, asleep.

These categories apply reasonably well to reports of strange sightings and interactions from all over the world. The modern era of such reports dates back to World War II, when stories of what we now call UFOs began to trickle in. Allied pilots described seeing "foo-fighters," luminous spheres that darted around Allied aircraft while on bombing missions, sometimes keeping station with the bombers and often undergoing extraordinary changes of course and speed. (We will explore this phenomenon in Chapter 7.)

The number of all types of alien phenomena reports grew rapidly after the widely publicized Kenneth Arnold sighting of UFOs in 1947—the sighting from which the term "flying saucers" was coined. The number of reports continues to grow and is now large enough that significant numbers of people have had an experience in one or another of the above categories.

Common elements of UFO sightings include, for example, certain shapes (cigars, saucers, wedges), certain attributes (silent running, for one) and out-of-this-world movements. Many UFOs move but reveal no details of shape or surface texture. Because they are seen by a wide variety of people and mostly for short periods of time in unforeseen circumstances, they may as a group contain a lower incidence of deliberate fraud than the second category—the sightings of what are claimed to be alien spacecraft.

The photos of alien ships produced in this second category are examined critically by specialists inside the UFO community, and are rejected publicly if found wanting (see, for example, *UFO Magazine* April/May 1999 or almost any recent issue). It is common for the images of spacecraft to be seen against a blank sky. The craft also generally fit into a few common categories of shape.

Because the third category, interaction with aliens, usually is freighted by heavy psychological baggage, the abduction reports may contain a high percentage of honest reporting, though the reality reported may be greatly at odds with common experience.

We will explore each of these three categories, but we will focus most intently on the first two. The reason is that, except for the reports of people passing directly through walls unscathed or of bloodless and incisionless operations (both of which are interesting as physical puzzles—see Chapter 6), the third category of reports generally omits behavior and technology that requires or defies *physical* explanation (although it may well require psychological explanation). We will mainly delve into the reports of UFO sightings

A 1952 close-up photo of a UFO taken over Passaic, New Jersey. Photo: copyright 2000 Fotorama/Fortean Picture Library

An American military plane saw a dark-red UFO at 15,000 feet over Utah and the pilot took this photo in 1966. No official explanation has been offered. Photo: copyright 2000 Fortean Picture Library

(category 1) and alien spacecraft (category 2), looking carefully at what the accounts imply or say about what happened, trying to show where conventional science leaves off, and then suggesting what kinds of new physics—alien technology if you will—might explain them.

These three categories are not exhaustive, and within them there is much variation as well as commonalities. Sometimes high-tech equipment is involved in the sightings. For example, on August 13, 1956, British radar operators at Bentwaters, England, observed a set of blips with intensities similar to that of an ordinary jet aircraft (before the age of stealth!). The blips were tracked as traveling at speeds of up to 9,000 mph, way above the capability of any maneuvering craft of human origin, then or now. This fits in the category of UFOs, but it wasn't a visual sighting.

In other cases of radar sightings, fighters have sometimes been scrambled to give chase to blips, and the fighter pilots have reported being unable to keep up with the turns and accelerations of the bogies that sometimes were directly visible to them.

The mass nighttime UFO sighting in Mexico described earlier has been repeated elsewhere. A similar one occurred in Arizona a few years ago. One unusual mass daytime sighting took place in downtown Mexico City and was viewed by thousands. A craft was seen to hover and move slowly above the buildings. I recently saw a video of the event, and the ship looks something like a blimp with hard edges and odd projections, and it appears to glide slowly behind several skyscrapers.

Occasionally, physical evidence is presented along with eyewitness accounts. A few cases of scorched earth have been put forward, in which plants died in patterns and locations coincident with shapes of UFOs, and where they were reported to land or hover. At some of those landing sites, indentations in the ground were seen, as if something heavy had rested there.

A second incident at Bentwaters represents a case with even more unusual physical evidence. Late in December 1980, security guards at a now-closed U.S. Air Force base saw unusual lights in Rendelsham Woods, just beyond the security perimeter. On the second night of activity, guards entered the forest with flood lights, geiger counters, and two-way radios. There they saw a craft they estimated to be 20 feet wide by 30 feet high. As the craft approached them their counters started to register counts faster than normal and their radios intermittently failed. A return to the site in daylight showed broken tree limbs and 7 one-inch-diameter, 1.5 inch-deep circular depressions. (A recent photo shows that the "dead patch" originally found in the center of the site is now green and the surroundings are brown. This is a curious turn of events for which no solid explanation is currently proposed.) A later British Ministry of Defense memo stated that soil samples taken from the site at the time showed levels of radioactivity 25 times that of the normal background. This is a case of a category 2 sighting with additional physical evidence.

In some abduction accounts, the abductees report that aliens have inserted an implant of some sort into them, their purpose not always known. In at least one of these cases, the "alien implant" was surgically removed and sent to a lab for analysis. The results were inconclusive. This is a category three case, but with a physical trace.

How should we treat these varied phenomena? To answer this question, it is useful to look at earlier examples of how we have thought about categories of unexplained phenomena observed in the natural world.

THE SKY IS FALLING: HOW WE HAVE TREATED UNEXPLAINED NATURAL PHENOMENA

Many tons of stony and metallic dust fall continuously into Earth's atmosphere every day. It is the debris of comets and the leftovers of asteroid collisions out between the orbits of Mars and Jupiter. We are unaware of this cosmic rain, except on annual "meteor shower" nights when it is particularly intense because the Earth is passing through a concentration of debris, much of which is only as large as grains of sand. If it is clear on such nights, we see a lot of "shooting stars." From ancient times up until the 1830s, most astronomers and philosophers thought meteors originated in Earth's atmosphere, along with comets. Western scientists didn't know they actually fell to the surface of the planet. It was only with the spectacular Leonid meteor shower of 1833 that scientists reached consensus that meteors originate from material orbiting the Sun.

You might think that we see inbound meteors more clearly now than in years past, given all our advances in technology. That is not entirely true. Though we now can observe meteors with many new instruments at wavelengths beyond human vision, the ancients' view of meteors and the sky as a whole was a good deal clearer than ours. That's because they had no bright haze from electric city lights, and they had much less air pollution than we have. It is likely that a higher percentage of people in pre-industrial times saw the meteor showers with better clarity than we do and were familiar with their appearance and range of behaviors. Furthermore, the known showers are slowly dying off with time, as their cometary sources continue to evaporate with each solar passage. The falling material also gets used up in the showers here and slowly disperses along its orbital path. Eventually these showers will die away. Even though these factors suggest that reports of impacts by meteorites would have been more common two thousand years ago than now, there really is little evidence in the historical record. The explanation for this is that such events are truly rare, and overall the population density was much lower than it is now, so the chance of seeing a meteor strike the ground was probably lower. However, a large meteor collected from Greenland by the American Museum of Natural History was known to the natives there before the arrival of Europeans, and the rock known as the Ka' bah in Mecca is a multi-ton meteorite deemed sacred by Muhammed 1400 years ago.

For a chunk of meteorite to strike a person when it crashes to Earth is (fortunately) a rare event that definitely would make the evening news. Yet, rare as this is, we believe meteorites do fall and that a few have even struck people. Why? Because we have, in at least some cases, a continuous chain of evidence about the event and also a model for how it happened. In other cases, we have partial accounts and find fragments in locations consistent with the observed trajectory, speed, and mass of the pieces. In just a handful of cases, the eyewitness was the target, and the damage done was visible on her or his body, as well as (in some cases) to the house, car, or even the bed they were in when it happened. In a number of these events, the remains of the meteorite were placed in the possession of museums and universities where they were analyzed. In many cases, pieces were lent out on demand to others for further examination, and some were eventually put on public display.

Even though the events are rare, there is verifiable evidence, including material evidence, and consistency with our ideas of how the world works.

Other rare events occur that seem consistent with the known world but have no simple explanation. Take the disaster of TWA Flight 800 several years ago. There were sketchy eyewitness accounts from people on the ground who said the plane exploded over Long Island Sound a few minutes after takeoff. A few people thought they saw a luminous streak cross the sky that intersected with the plane just before the explosion. This was never confirmed, and no one was able to prove that any missile-launching aircraft were nearby. The "black box" recorders which held records of the last seconds of the flight, most of the pieces of the wreckage, and bodies were recovered from the plane and carefully examined. The consensus at the FAA is that an electrical spark in a mostly empty fuel tank set off the explosion, but there is no hard evidence to prove it. This is the hypothesis left, consistent with most of the facts, after several others (such as a bomb on board or an impact by an air-to-air missile or meteor) were shown to be wrong.

There are of course also *natural* phenomena that are not well understood, but they do not seem to violate any known laws of physics. Consider the following example: Ball lightning has been seen by people in many different circumstances. It has the appearance of a luminous blue-white fuzzy sphere that travels along electrical conductors and often generates a spitting-buzzing sound, and it can cause static on nearby radios. From the above circumstances it seems clear that it is electrical in nature. It is probably a self-contained low-temperature plasma—electrically charged gas—which has an internal magnetic field that helps it to be relatively stable. However, it is not clear how it starts or how it maintains itself. More than one reported UFO sighting has turned out to be an incident of ball lightning.

For another example, consider the seasonal migrations of birds, butterflies, or whales. They travel thousands of miles to return to their breeding grounds with great reliability. How do they do it? In some cases, there is evidence that magnetic particles in their brains work as a compass. In others it seems that star patterns allow a sort of celestial navigation. In the case of the whales, perhaps the patterns of currents and of temperature variations provide the clues. But how do the whales remember these in the required detail? We don't know. These phenomena may be hard to explain but seem unlikely to violate known laws of physics.

All of the cases of unexplained natural phenomena cited above fit into the first category of UFO/ET phenomena. That's because for years they were (or still are) unexplained, and physical evidence was often lacking. They did not obviously violate known physical laws (some UFOs do, however), and later some were explained and fit into the known laws of physics. Perhaps someday UFOs will enter the ranks of the understood.

SCIENCE OVERTHROWN: NEW PARADIGMS

This is not, however, the end of the story. If it were, science would be a dead discipline. At some point, repeatable observations differ from the predictions of theory, and then science advances, often by a great leap. At the end of the last century, two observations were made that seemed completely unconnected at the time, though both violated the then-current physical theory. First was the discovery of the change in orientation of the orbit of the planet Mercury. Its orbit is elliptical, like the rest of the planets in our solar system. But this orbit was observed in the 1700s and 1800s to be slowly changing its orientation in space.

Careful calculations by astronomers such as Simon Newcomb showed that Newton's physics did allow for the perceived changes in Mercury's orbit. In effect, Mercury travels an ellipse with a slightly different position in space on each and every orbit, due to the equatorial bulge of the sun that exerts an asymmetrical force on the planet. The calculations based on Newton agree qualitatively with the observations. But the calculated rate is only *half* of the observed rate. Though the observations and theory were reexamined by a number of prominent astronomers in the late 1800s, no error or explanation could be found for the discrepancy. It would take a revolution in our understanding of the nature of the universe before the puzzle would be solved.

Second, in the early 1880s, Albert A. Michelson developed a technique using light's wave property of being able to interfere with itself, to measure the speed of light to a precision of a few parts per million. He and a colleague, Edward W. Morley, expected to find, as Newtonian physics predicted, that the speed of light would depend on the direction from which the light was traveling through the universe.

Up to that time the majority of scientific opinion was that light needed a medium to travel in, as sound waves propagate through air. The supposed medium for light was called the "ether." As light traveled through the ether to Earth, and Earth revolved around the Sun, the direction light waves traveled inside Michelson and Morley's experimental apparatus would change with respect to the ether, if it existed. Thus the speed should change, too, just as the speed of a boat's wake depends on the boat's speed through the water. To their surprise Michelson and Morley found no change in the speed of light at all no matter what season or time of day. This result seemed to say that light did not travel through a medium and, therefore, there was no ether.

The failure to find the ether remained another unsolved puzzle until twenty years later when Albert Einstein used Michelson and Morley's result as a major foundation of his Special Theory of Relativity: the speed of light in a vacuum is an absolute number, unlike anything else, and is independent of the motion of the source and of the observer who measures it. No matter their own motions, observers will always find the same result. The discrepancy in the observed and predicted rates of change in Mercury's orbit was the first observational result—inexplicable by Newtonian physics—that Einstein could explain with his General Theory of Relativity.

The implications of relativity theory were revolutionary and flew in the face of classical physics in many respects. For some years, many classically trained physicists opposed the theory and its implications. It predicted things that were outside common experience, and it seemed to violate common sense. Yet, it prevailed because it had the key elements of classic, solid science: a self-consistent theory that was able to make predictions of new phenomena; the agreement with established results to a higher precision than any other theory; new observations continued to confirm the theory and every repetition of them with better technology has brought better agreement. These, and some curious predictions of quantum theory about instantaneous communication, will be the subject of Chapter 4.

The implications of relativity theory for space travel in ships include such things as limitations on the ultimate speed of travel and the stretching of time and the slowing of all clocks, including biological ones, as you approach the speed of light. We will look at these areas in Chapter 3 and see how they relate to UFO phenomena. Relativity also holds the theoretical possibility of traveling through enormous distances across the galaxy in very little time, if not instantaneously, via space warps and holes in spacetime.

"HE PULLED A RHINO OUT OF A HAT!": THE WEIGHT OF EYEWITNESS ACCOUNTS

To date, just about all UFO sightings and stories of alien abductions rest on eyewitness accounts. The other elements of understood phenomena are usually missing. There is no self-consistent model to test against or make predictions from because in most cases the witnesses are not trained scientists, and they have little scholarly basis for generating theories about what they saw. No blame attaches to them because of that, and occasionally technically trained personnel see something, too. In some sightings, the witnesses report phenomena that appear obviously (in some cases subtly) to violate known laws of physics and common sense.

There is generally no unambiguous physical evidence nor a real chain of evidence, and no aliens have been produced for scientific and public inspection on demand. Reports by reporters without training and phenomena without precedent resist interpretation, and in some cases these even make speculation difficult. This makes it difficult for the technical community to believe any UFO account. In fact, for just about all knowledgeable scientists these flaws are fatal.

But they need not be fatal for us. We can still ask how much weight to give to eyewitness accounts, and also what we can learn from them, even if what we learn reveals more about human foibles than about alien technology. We can examine how incomplete evidence is generally handled in other human spheres of action and how we can make the most of what information has been reported. We will do this in the remainder of this chapter.

And in following chapters we will also look carefully at how to explain some of the more difficult to understand reports using both accepted scientific theory and leading edge speculation.

First, because eyewitness evidence is most of what we have to work with in examining UFO phenomena, we can certainly ask how eyewitness accounts are treated in other circumstances. There may be lessons to be learned by the group of interested lay people and scientists who collectively comprise what I call the UFO community.

Some lessons have already been learned. Starting in the late 1960s, after years of disorganized publications of annecdotal UFO reports and self-published books about contact with aliens, members of the UFO community reacted to the steady hand of astrophysicist Dr. J. Allen Hynek, who consulted on the massive study and compendium of UFO reports by the U.S. Air Force, called *Project Blue Book*. People in the UFO community began to form organizations to systematize the intake of reports, to try to improve the information garnered in each, and to attempt to analyze them. Project Mufon (Mutual UFO Network) is an early example. Other more recently formed groups include JAHCUS (the J. Allen Hynek Center for UFO Studies), UKPRA (United Kingdom Phenomenon Research Association), and those

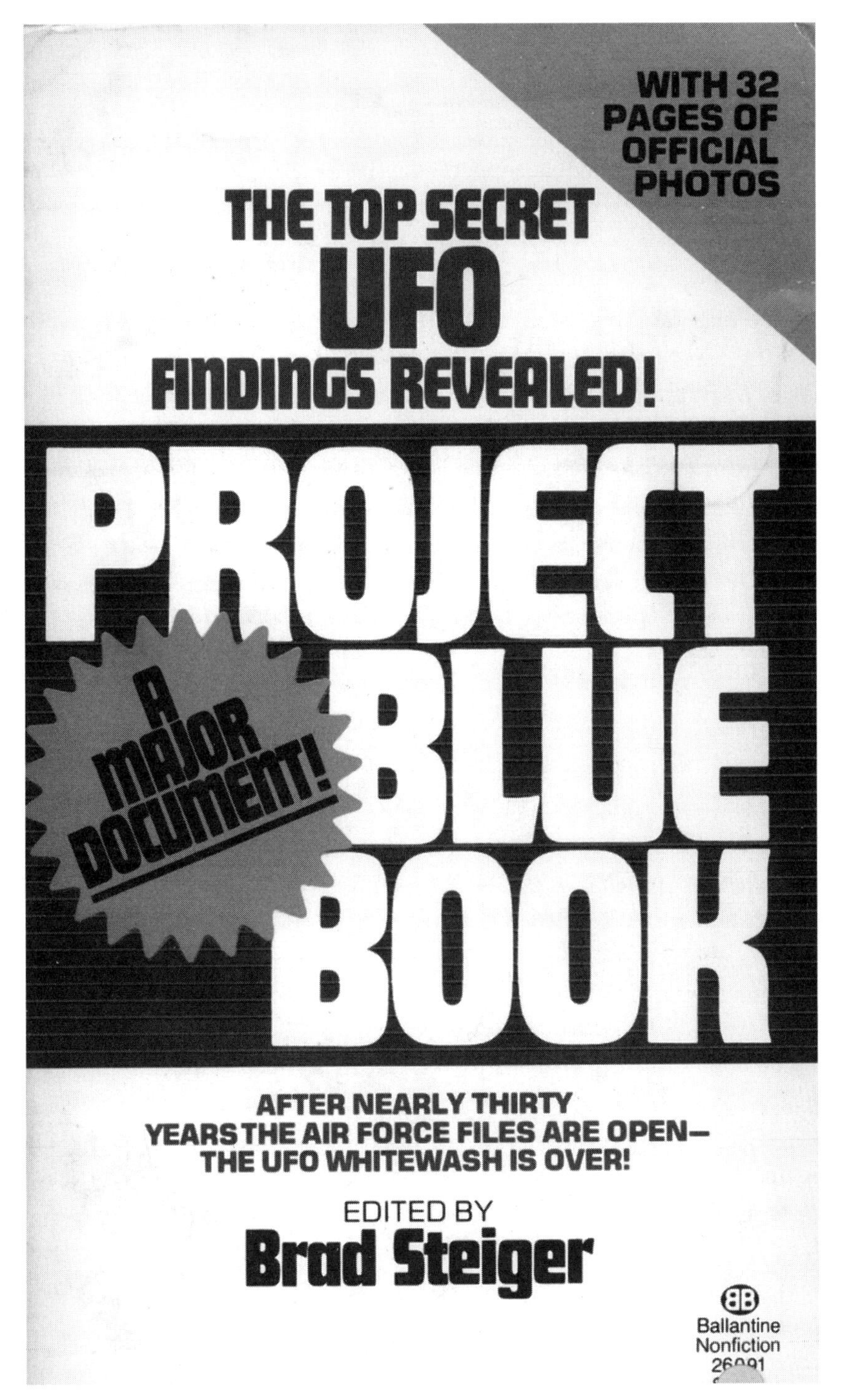

The government's famous Project Blue Book did little to clarify whether or not UFOs represented a real physical phenomenon. Cover photo: copyright 1976 Ballantine Books

involved in the research that supports the British publication *UFO Magazine*, to name a few. In the latter publication the discussion of incidents is reasonably coherent, and controversies within the UFO community are presented as forthrightly as those between UFOers and the technical community. Another example may be found in Bertil Kuhlemann's article, "40 Years of UFOLOGY in Sweden" *(MUFON 1987 International UFO Symposium)*. He showed a set of thirty-five standardized questions for data gathering, and has applied them to about 1,000 cases collected in Sweden over the forty-year period from 1947 to 1987. This is a step in the right direction. Besides the advance in data collection, this project has collected enough samples to allow meaningful statistical analysis to begin. In Chapter 7, I will discuss the history of UFO reports and cover some of the statistical results in more detail.

THE COURTS OF LAW AND SCIENTIFIC LAW

Let's pick two mundane examples for comparison to the UFO accounts: law enforcement and natural science. Law enforcement agents generally have less control over gathered evidence than scientists do, and in that respect law enforcement is more comparable to the current efforts of the UFO community. Scientists usually enjoy the luxuries of control over observation and experiment, of time to plan and contemplate, and even more important, of repeating an experiment in the same or other ways to allow confirmation or refutation. So far, those studying UFOs have had no such opportunity.

In law enforcement, eyewitness evidence is frequently employed to provide evidence of guilt and, occasionally, innocence. The prosecution will use eyewitnesses to try to prove the guilt of the accused. A single "good" identification is valuable, and "good" multiple identifications are even better. But what is the meaning of "good" in this case? First, the witness must appear to be certain. The ability to give detailed testimony that includes unpublicized details only known to the police is helpful. A witness who has the appearance of reasonable intelligence and coherence under friendly and hostile questioning lends credibility to his or her testimony. A witness is even more creditable when his statements are obviously true, don't violate common sense, can't be shown to violate laws of physics, or aren't refuted by other evidence.

Though perhaps it shouldn't, the social status of the witness (if he or she holds a "good job" or is a "pillar of the community") can positively reinforce the testimony in legal proceedings, and the appearance of unreliability can affect it negatively.

Finally, the witness should demonstrably have no "ax to grind." That is, he or she should have no relation to the victims, to the accused or his associates, or law enforcement officials. He or she should not be prejudiced by an ideological position, nor should he or she be under a threat by the accused. In many cases this last set of criteria is hard to achieve, simply because in the real world eyewitnesses are often directly involved in the circumstances of the crime, and they may have reasons to lie.

The social status of the witnesses who commonly report UFO sightings has, I believe, affected the view of those who are skeptical of such reports in two different ways, the first more obvious than the second. First, if the person reporting was said to be tired or under the influence of drugs or alcohol, is a social "drop-out," very poor, homeless, or mentally impaired in any way, their report will almost certainly be discounted. You can see in the serious UFO periodicals (those that attempt to apply screening criteria of a rational sort to the reports they examine) a fairly strong skepticism of such accounts.

The second bias is more subtle and may affect society as a whole. The vast majority of UFO sightings have been recorded outside of cities even though the world's population mostly lives in cities. There are several reasons for this. In most cities the streetscape fully occupies the attention of the inhabitants visually. They have to watch their surroundings as a matter of survival and cities are also visually interesting and complex to most people. As a matter of course most city people don't look up at the sky as they travel from place to place. At night the city lights are mesmerizing and will also blind pedestrians to anything in the sky that is fainter than the brightest stars and planets, so "extra" lights in the sky pass unnoticed unless extraordinarily intense. The noise level in cities is high and masks any unusual noises, such as the passage of a UFO (though most are reported to be soundless). There is also the possibility that, if there are actual alien visitors, they tend to avoid cities for their own reasons (but note the exceptional case of Mexico City cited above).

The net result is that because most UFO accounts are recorded by rural citizens, and because there exists a near-universal stereotype that country folk are unsophisticated and thus easily fooled or mistaken compared to urban dwellers, there is likely an unconscious prejudice against their stories. I am probably guilty of this myself.

In thinking about these problems, it should be kept in mind that UFO reports have been made not only by untrained observers of the countryside, such as farmers and "good ol' boys" out in the fields and bayous, but also by engineers, scientists, pilots, technicians, and others who have technical, naturalist, and scientific training, and who are by virtue of that training careful observers. If we consider the anti-country bias on its own, we should admit that by virtue of their living in the country where life is slower and quieter, the horizons unrestricted by buildings and the night sky darker, and also because farmers, lumberjacks, military and country police personnel often closely observe their surroundings as part of their work, they might be better observers than those of us who live in cities.

EATING HUMBLE PIE: THE AUTHOR'S UFO SIGHTINGS

Of course, though I have been a dweller in cities for many years (Chicago, Boston, New York, and now San Francisco and Los Angeles), I am one exception to the rule in that I have had my own city sightings. I have thought of myself as an astronomer from childhood. I received a doctorate in astronomy, and though I am not currently doing research in astronomy, I teach it and I look at the skies everywhere and often. I enjoy spotting the planets, the Moon and constellations, atmospheric phenomena such as rainbows, sunsets, rings around the Moon and sundogs, and human events such as jet contrails, searchlight beams, and night sightings of illuminated blimps. I observe the sky wherever I go.

I have had two UFO sightings myself, and they both took place in a city. I will describe these here because they are examples of how a trained observer can be unable to explain what he sees.

Both incidents occurred when I was home from college for the summer and held a job at the Adler Planetarium in Chicago. At the time, I was an astronomy major with long and varied observing experience with both amateur and professional telescopes. My job at the planetarium included answering questions from the general public about all kinds of things, including exotic cosmologies and flying saucers. I had gone through a stage in grammar school of believing intensely in the existence of flying saucers. By the middle of high school I had read George Adamski's books about Venusian visitations and a lot of science fiction, and had then with my increasing age and knowledge decided that most of the UFO reports to date had been balderdash. I also thought I "knew everything" about the skies.

One evening, however, I walked out of the planetarium to go home just as dusk was falling, and as usual I took a slow look around the horizon and sky. The planetarium is on a beautiful site on the shore of Lake Michigan, next to Meigs Field (for small planes) and just east of the Field Museum of Natural History. From there the lake stretches out to the eastern horizon, but to the west there is a wonderful view of the whole Chicago waterfront, with beautiful parks and spectacular modern skyscrapers. As I looked southeast over the lake, I saw to my surprise a disk, soft white in color, smooth and unshadowed, about the size of the full moon, hanging silently and motionless about 50 degrees above the horizon. I had never seen anything like it, and I stood there for about five minutes mentally running through every explanation I could think of.

I was stumped, and then it hit me that all of Chicago could see it. If experience was any guide, I knew the planetarium switchboard would be lit up about it, and someone on staff would have investigated and found an explanation. Feeling rather sheepish about reporting something that, if I were wrong, would perhaps lower my status in the eyes of the staff astronomers, I went back inside and found I was correct. Calls were coming in and someone had checked and found that the object was an airborne cosmic ray experiment with a detector slung under a large balloon. It was far enough away and the light poor enough that no details of the detector basket or balloon shrouds were visible to the naked eye. All you could see was the plain white balloon. Over a period of several hours the winds shifted and it slowly drifted away. It was recovered days later in Texas.

The second sighting came later the same summer. I was in a park just after dark, about 9 P.M., looking at the Moon with my telescope. It was a balmy, clear night, and my southern horizon was lit up as usual by the scattered lights of Chicago. As I looked across the western sky, out of the corner of my eye I saw to the south an undulating ribbon of white light, which varied slightly and irregularly in brightness as it moved slowly east. Its length was about one quarter of a degree—one half the diameter of the full Moon. With some difficulty I pointed my telescope at it and followed it as it moved. At a magnification of about 50 power, its edges, though sharply defined, wavered like a banner blowing in a breeze. It was silent and left no trail. After a moment's thought, and based partly on my recent experience, I went inside and called the Adler Planetarium. Again they had the answer: it was a flying advertisement being dragged behind a small plane, and it was lettered using many light bulbs. Even with my telescopes, the distance was too great (twenty-five miles away, over Chicago) to see the airplane, and the angle was so low that I viewed the ad nearly edge on, so all sense of the lettering was lost. To this day, I think of this sighting as my personal view of Quetzalcoatl, the "sky serpent" of ancient Mayan religion.

These two experiences were lessons in humility for me. They taught me that not only do I not know everything but I should also have sympathy for untrained observers who are faced with unusual sights they can not explain. Even trained observers can report phenomena accurately, yet have no idea of their exact nature. I also believe that these two cases demonstrate that in some ordinary circumstances there may be no way to deduce from the evidence of an unusual sighting what was actually happening. Only specific "outside" knowledge supplied the answer. With the additional knowledge, these phenomena became easy to understand as part of our ordinary world. Neither required one to believe that aliens were involved. Yet, before the answer to the puzzle was found, one could easily have analyzed these cases for alien capabilities. Because we are missing that outside information in unexplained cases of UFO reports and related phenomena, we will make our best guess as to what physical capabilities and mechanisms might be involved assuming they *are* alien to Earth, and also try earthly explanations if any suggest themselves.

MY OX OR YOURS? WHO GAINS?

The issue of who can gain from making UFO reports is of major importance. The question needs to be asked because in the modern world everything can be put up for sale in one way or another. An industry exists in the sale of materials and tourist sites related to UFOs (have you visited Roswell?). UFO believers sometimes claim that accounts must be true because the witnesses have nothing to gain and perhaps a lot to lose by coming forward with what might on the surface seem to be an unlikely or even crazy story. This is a negative proof and one that is hard to refute in some cases. However, there is no published information I have seen that shows what percentage of people who reported their UFO experiences lost their jobs or social status as a result.

Those who sell videos and books may make money and achieve at least limited fame, although there is indeed a downside. Author Whitley Strieber wrote several best-selling books (*Communion*, et al.) based on his self-confessed alien abduction experiences. His books not only generated significant cash flow but the incredulity and derision of his peers in the publishing world. He must now either continue to insist that his stories are fact-based, which puts him on the literary fringe to say the least, or confess that they are fiction, which will lose him his fan base.

Having said all of the above, I must say that eyewitness accounts of all sorts are notoriously unreliable even in the best circumstances. Careful observers can be fooled about what they believe they saw. Sometimes, it's a simple and honest error, sometimes it is due to subtle "coaching" created by the mode of questioning by law enforcement officials or others who, in effect, "lead the witness." A particularly sad example of the latter can be found in a number of trials of daycare center workers for child abuse that occurred in the 1970s and 1980s. In many of these cases, reviews of the evidence, which was principally videotaped testimony of the children involved, revealed that the questioners, who were often supposedly well-trained child psychologists and juvenile-crime-unit detectives, had "led" the witnesses. They also conducted extremely long sessions that wore the children down. The latter technique is also used against adult crime suspects, and history shows this can and has led to false confessions of guilt.

Prosecution witnesses sometimes make statements that they cannot actually support, though they are "morally certain" that the accused is guilty. Occasionally, the actual perpetrator of a crime and another person look much alike, leading to mistaken identification. In some cases, personal prejudice and ethnic and racial stereotyping play a role in the error.

Any competent defense attorney will attack the prosecution's eyewitness testimony on any of these grounds, if he or she can. The defense will also call alibi eyewitnesses if it can, which the prosecution may attack on all of the same grounds.

In any case, eyewitness testimony gains credibility if there is independent physical evidence that is consistent with the testimony. Even though "good" eyewitness testimony may enable a jury to say they believe in the accused's guilt "beyond a reasonable doubt," and this is sufficient to convict, in many cases such testimony has proved insufficient in the absence of physical evidence. The highly publicized case of Rubin "Hurricane" Carter is an example of the poor use of eyewitness testimony and the importance of physical evidence. In the court of law, cases can be decided on the basis of eyewitness testimony without physical evidence. In the arena of the sciences, however, the criteria are different.

GETTING DOWN WITH SCIENTISTS

Scientists always want physical evidence to use in any discussion of new phenomena. Let us examine several examples. First, we have already described above the chain of evidence that eventually led to our current understanding of meteors, in spite of the early view that they were of Earth and not from beyond our planet.

Second, consider the narwhal. For centuries, sailors reported seeing a beast with a single straight horn on its head, and sometimes they had the horns themselves to prove it. Sailors sometimes said the beasts looked like unicorns, which the sailors claimed to have spotted on exotic shores. Eventually, live whales with single straight horns on their foreheads were spotted by competent naturalists, and that explained the sailors' eyewitness accounts.

Even stranger was the case of the coelacanth. In 1938 an amateur naturalist visiting India heard rumors of a peculiar fish caught in the Indian

Ocean near Madagascar. She eventually tracked down one in a market. She described it to a professional paleontologist who tentatively suggested that it belonged to a genus of fish thought to be extinct since before the end of the age of dinosaurs. He was quite skeptical. But it turned out that these fish had been taken sporadically for years. Eventually, several more specimens were bought from fishermen and examined by professionals. They dissected the fish and found the hand- and foot-like fin bones and eye socket details characteristic of the primitive coelacanths, well known from fossils more than 70 million years old.

A few years ago, rumors circulated that a strange species of deer had been spotted in the jungles of Laos. After a number of attempts, and despite serious skepticism about the existence of an unknown species of large animal that had previously escaped detection, it was seen and photographed. These animals are quite shy and clearly rare, are an endangered species and have been given protected status as a result.

Again, in each of the above cases, rare and unexpected observations drew initial scientific skepticism, but they were later shown to be correct and understandable through existing scientific knowledge. We have not yet reached the point of understanding UFOs, but perhaps some day we will.

"ANALYZE THIS!" THE AVAILABILITY OF PHYSICAL EVIDENCE; THE KINDS OF TESTS WE COULD MAKE ON PHYSICAL EVIDENCE TO PROVE A POINT OR DISPROVE IT

Based on all these examples and arguments, it might seem that physical evidence and multiple sightings would be key to the acceptance of alien visitations. Of course that is true, but will a simple submission of, say, an alien ice bucket settle the case? Suppose that one day someone steps forward to claim contact with aliens and offers not only an eyewitness account but also physical evidence, including some nonterrestrial tools. How would the world know it was real and not a clever hoax? After all, there are a lot of talented people in the entertainment industry who are masters of makeup and special effects. What sorts of tests might we perform to evaluate the truth and what kinds of evidence would be convincing? That depends on the kind of evidence being offered.

First, there are certain artifacts and accounts from the ancient world that are often associated with aliens and the arrival of UFOs or alien technologies. The materials are there on the ground staring us in the face, but for many people the construction technique seems a total mystery. This class of phenomena includes the Egyptian pyramids, the giant stone fortresses erected by the Inca, and the huge statues on Easter Island, among others. We will look at ancient alien visitation possibilities in Chapter 7.

Second, there are the artifacts made of "substances unknown to human science," often described in classics of the golden age of science fiction. In these stories, authors wrote about unknown alloys of known elements, a clever extrapolation of known science that did not strain the credulity of the reader unduly. How would we identify such an object if it were handed to us? The most obvious way is if it had properties at odds with all known materials. It could have an appearance unlike any seen before. For example, it might be a perfect black substance absorbing light from all directions. Even the best black coatings we have reflect a small amount of light. A perfect isotropic black (black at every angle of view) would have the unfathomable appearance of a black hole.

A space suit material that is completely airtight yet entirely flexible—and thin enough to be used for surgical gloves—would not be duplicable by current terrestrial technology. In fact, space suit gloves and space suit joints in general are the toughest part of human design for space travel at the moment (with the possible exception of zero-g toilets).

A material that has a strength-to-weight-ratio exceeding that of silk or bamboo or steel would be of interest. There is a good likelihood that cylinders of buckminster fullerene, also called buckytubes, exhibit such strength. They are made of a recently discovered form of pure carbon with atoms arranged in a regular polygonal pattern, like the intersections of chicken wire filaments or the pattern on some caned-seat chairs and woven baskets. So far buckytubes are a lab curiosity, and no working examples exist outside the laboratory. All tests have been done with microscopic quantities.

One aspect of the famous Roswell, New Mexico, alien spaceship crash accounts is an often-made claim in precisely this area: that the very thin metallic remains that were recovered at the site were of super strength, beyond any known material made on Earth. The published photographs taken in 1947 show a thin metal foil that has been ripped into pieces smaller than a desk top, so that material did not have infinite strength, and it looks like ordinary aluminum foil. But it has been alleged by eyewitnesses from Roswell that the actual crash-recovered material was not shown to the public or press.

If we could manufacture materials that conduct electricity with zero resistance, we could save about 15 percent of all the electricity we generate that is now lost from the transmission lines. We could perhaps banish forever the possibility of fires from overheated electrical circuits. Such superconductors that could operate at room temperature are a widespread dream. They have been sought after seriously since the discovery in the 1970s of materials that become superconducting at temperatures more than a hundred degrees above the old limit (4 degrees above absolute zero). The consensus is that such materials are at least twenty years away, so a demonstration of one now would be a reasonably convincing sign of non-terrestrial origin. The flip side of superconductors is perfect insulators. Because they don't exist here they would be convincing artifacts of alien technology.

Lubricants that yield zero friction in a conventional bearing or on conventional surfaces would be incredibly useful but don't yet exist. They would extend the life of almost any mechanical device with moving parts.

In all of these scenarios, there is technology humans haven't achieved yet—but they possess no obvious contradiction of known physical law. The elements of their composition might all be known to us, but presumably upon a chemical analysis the combinations would be found to be unique and perhaps even evade some "known rules" of chemistry. In other words, these would be alloys or composite materials unknown to humans.

It is also possible that such non-terrestrial artifacts could be composed of new elements, as yet undiscovered by humans, not just alloys of known ones. The periodic table includes all elements so far discovered in nature, from hydrogen through uranium, plus a large number of unstable isotopes of these and a few heavy elements that are unstable in every isotope. (It was Mendeleev's mid-1800s success at predicting the material properties of elements missing from the periodic table that led to the later adoption of his version.) Many of these isotopes decay radioactively in a fraction of a second and, even though they may be created by natural processes, disappear so quickly that the only way to discover and study them is to generate them in a subatomic particle accelerator.

There might be a whole family of elements, not yet discovered by humans, that are super heavy. Any object made of an element above atomic number 140 on the periodic table, for example, would gain immediate legitimacy as an artifact of alien technology. All these possibilities will be examined in more detail in Chapter 8.

When we leave the realm of products and materials that do not contradict known physics and therefore could be said to be beyond us at present but are still foreseeable, we're basically entering the realm of science fictional concepts. This doesn't mean the "world of the possible" has been left behind. Many modern marvels, from radar to communications satellites and cell phones, were predicted and first explained to the public in science fiction stories and novels.

HOT CONTROVERSY, COLD FUSION

Before we head for alien realms, let us consider a recent case in which reputable scientists made a claim that appeared to contradict known physics: The case of cold fusion. Think of the end of that delightful movie *Back to the Future*, when Doc (Christopher Lloyd) drives up to the house of Marty McFly (Michael J. Fox) and feeds veggies and garbage into a unit that looks like a Cuisinart® and is described as the "fusion generator." High temperature controlled fusion has been the Holy Grail of energy research since the 1950s, and even though there has been large funding for the program in the United States and a number of other countries, a practical reactor is probably decades away.

In the 1980s a scientist at a university in Utah claimed that he had found evidence of nuclear fusion energy generation in what looked like chemical reactions at room temperature. Then a young electrochemist at another Utah university, who was teamed up with a much older and world-renowned electrochemist from England, announced they had seen evidence of fusion in a different reaction, one also at room temperature in a desktop-size apparatus they had designed and built.

No theory existed to explain any fusion that runs at room temperature. A small part of the scientific community both here and abroad rushed to duplicate their results and concoct a theory to explain them. It was announced that the university had submitted patent applications and it anticipated becoming extraordinarily rich by licensing the technology. But the results were non-reproducible, and with the exception of some diehards the idea has now been dropped. The vast majority of scientists were extremely skeptical of cold fusion precisely because it violated known physics.

Now on to even more exotic stuff.

A material or finish that changes its appearance, including texture, color, pattern, or reflectivity, either on command or automatically, depending on internal logic or external sensors, or an object that changes its shape to conform to the hand that holds it, would create a sensation. Physical shape-changing without expenditure of energy would violate what we know about changes between states of matter and conservation of energy.

A material that exhibits antigravity and, even better, switchable antigravity without expenditure of energy, will be convincing as a product of alien technology. Nothing like it exists on Earth. While both Newton and Einstein seem to allow for natural antigravity in their equations, neither predicts it and neither permits switchable gravity or antigravity (no natural antigravitational matter has yet been found). Our understanding of physics would also seem to rule out gravitational shielding, the existence of which would perhaps explain the silent lift-off of spaceships mentioned in many alien craft sightings. However, it should be noted that, though it is not known to the general public, Einstein suggested there might be a gravitomagnetic force of some sort. Because magnetism is the handmaiden of electric current—electricity is controllable, and both magnetism and electricity come in two polarities (north/south, positive/negative)—it is perhaps possible that this concept could someday lead to controllable antigravity.

A trick related to and almost equivalent to the above would be the ability to cancel inertia, a fundamental property of matter. Inertia is the tendency of matter to resist changes in its motion or state of rest. Inertia and conservation of momentum are what press you into your car seat when you put your foot on the gas. They are also what would turn you into a grease stain on your commander's couch if your starship did not have inertia control and you tried to accelerate quickly from rest to light speed. If aliens arrive in a sub-light ship with only one-g acceleration (note: one g for *their* planet), inertial control is obviously not an issue. For any warp or near-light-speed drive that accelerates at more than 10 g's for more than a few seconds, inertial control becomes a serious issue. All warp speed ships on *Star Trek* have "inertial dampers" for just this purpose.

A ship or object that operates via a new force, a force other than the four fundamental forces we know of (gravity, electromagnetism, the strong and weak nuclear forces), would be spectacular proof of alien origin. One obvious candidate is the gravitomagnetic force mentioned just above. Combinations of the other forces would also do. (Physicists believe the four forces we know about were combined at the very beginning of the universe, in the first fractions of a second of the Big Bang.)

An alien in possession of alleged wormhole technology could demonstrate its validity first by showing an instant view of the cloud system of Jupiter and the state of volcanic eruptions on its bright moon, Io. This could be a conclusive (and colorful) test because, depending on where the Earth and Jupiter are in their orbits, the light that would show us the state of the clouds and volcanoes at test time would take between about 25 and 50 minutes to reach us, traveling at the usual 186,000 miles per second. We could call the first view "instant preplay!"

A second and more strenuous test would be for the aliens to take some scientists, along with their instruments, to Mars in an instant. Our people would, of course, need to take spacesuits if they wanted to walk around on the planet. A convenient and convincing version of this would take the form of a hole in space that you could simply walk through, stepping onto another planet or even just across the continent on Earth. This was envisioned in the classic novel *Tunnel in the Sky* by Robert Heinlein and visualized on the current TV show *Stargate SG.1*.

A different version of this ability to jump light seconds or light years is the matter transmitter. Most versions of this in fiction suppose that it is possible to disassemble a thing, living or not, and transmit it instantly, or at least its pattern, in perfect detail to the destination ("Beam me up, Scotty!"). If the equipment were set up at different ends of a lab and say a rabbit were put in one compartment and it disappeared there and reappeared at the other end, that would be a sufficient demonstration of the principle. To demonstrate the instantaneous long distance capability, you could send someone to the surface of Europa, another of Jupiter's moons, equipped with a space suit and a powerful laser. She would point the laser at Earth and let it give out a set of coded flashes. Then she would step back into the transmitter and beam right back. Confirmation would come when earthly observers note the laser flashes about 25 to 50 minutes later.

If you can transmit matter you can also almost certainly synthesize it. Matter synthesizers could be tested by providing objects to be duplicated and test the results for exactness of duplication. It is possible that biological beings could be instantly cloned using a synthesizer and, of course, a DNA test would be able to check this. This would be an easy way to have unlimited chocolate milkshakes on demand, or to overpopulate the world with your own clones. Restraint would be called for!

An even more convincing display of alien technology would be a force that is none of the four known ones—an entirely new spectrum with force laws entirely different. What do the phrases "new spectrum" and "new force law" mean? A new spectrum implies a particle or wave of a new type that has some property other than the usual ones of mass, electric charge, spin, etc. A new force law means that the strength of the force varies with the distance between these new particles at a rate different from the four known forces. Physicists would have to perform experiments to eliminate known possibilities, and these would involve multiple trials with test objects of different types of matter, charged and uncharged, magnetic and not, etc., to establish the facts.

"DAVE, I'M FRIGHTENED": THINKING MACHINES AND ALIEN BIOLOGY

A truly intelligent computer would also be a worthy candidate for an example of alien technology. Many research and development programs are working on creating one, but so far no one has come close. Alan Turing, the British pioneer in computers, invented the principles of stored general programs and helped the Allies crack the German Enigma codes by using machines built on his principles. He proposed a test for machine intelligence generally accepted today: if a human communicates with another "entity" not visible to him, and by communicating cannot tell if the other thing is human or machine, then if it *is* a machine it is intelligent. The computer-generated TV character of the 1970s, Max Headroom, was a clever and convincing portrayal of such a thing. The character HAL, the sentient computer in the movie *2001: A Space Odyssey,* was another. *The Forbin Project* computers, Colossus and Guardian, were able to hold an intelligent conversation and would probably also qualify.

There is still a great deal of controversy over whether a truly intelligent machine is possible. Roger Penrose, one of the twentieth century's leading mathematicians and cosmologists and a colleague of Stephen Hawking's, in his book *The Emperor's New Mind*, argues that it is impossible. He suggests that the human mind, and perhaps all organic minds, are by nature quantum mechanical, and that no artificial mind can have this property, thus none will ever be intelligent. I do not agree. I believe that quantum mechanics plays a role only in the behavior of molecules and atoms in the brain, and that the brain's processing capabilities arise out of the group or system behavior of all the neurons, and thus the numbers of them and the numbers and types of their interconnections are what determine their abilities. The evolutionary development of this behavior is not yet understood, but there is a name for it: emergent behavior. So I believe it is only a matter of time before we succeed in making intelligent machines, perhaps simply by making an electronic version of the brain with a lot more interconnections than exists now. A candidate alien who could present a lively artificial conversationalist, or who was one himself, would be convincing.

A device that stores more information by a factor of say ten thousand than any of our current devices, or allows instant access to a "galactic library" that is the sum of all known knowledge from intelligent space, with full color and motion views of other worlds, and of live events anywhere on Earth, would be beyond us and thus alien in origin.

ALIEN PSORIASIS

It is easy from here to hop over to biology. The possibilities for the biology of the aliens themselves, and any plants or animals they might bring along (either as food or living examples), are legion. But if their flesh or circulatory fluid can be sampled and tested, their unique biological code should be easy to find, with one exception. That is the one in which the aliens are identical to us, for whatever reason. However, even if we are identical, I believe it unlikely the aliens' plants and other animals would be, too. The first reason is the workings of evolution. If aliens "planted" our forebears on Earth, it is almost certainly a minimum of 30,000 years since that event, based on the current oldest archeological evidence. Let us assume that this happened either via starship or matter transmitter. The intervening time, though it is short on a geological time scale, is still long enough for the plants, animals and especially any bacteria and viruses to have evolved away from one another on the two planets. This would happen because the conditions on Earth would have been significantly different from the aliens' (and therefore our) "home planet" for that whole time. Different ecologies breed different species. And simpler organisms breed faster and evolve faster than complex ones. The second reason is that if the aliens have star travel they also almost certainly will have genetic engineering far in advance of ours, and they will have used it to make new species. Again, these side dishes will be different from ours.

These "identical" aliens might demonstrate their off-planet origin by eating foods we don't have, breathing an atmosphere we find poisonous, or showing that ours is deadly or damaging to them. (Think of the chlorine breathing aliens of many sci-fi stories.) If they are akin to us, it is most likely they will breathe oxygen, exhale carbon dioxide, and have a carbon-based chemistry. It is unlikely they will be based on anything other than carbon. Silicon, the most likely substitute, is more plentiful than carbon on Earth, yet not one documented silicon-based organism has ever been discovered here.

If our extraterrestrial visitors are truly alien, they will not look like us. The most obvious difference will be morphological: the aliens will likely look like no known species. (Keep in mind the limits to our knowledge of terrestrial species discussed above and the estimate that there are still about 6 million *uncataloged* species of insects here on Earth.) They will likely be bilaterally symmetrical; evolution here has favored that as a general, though by no means inviolable, principle in the phyla of complex organisms. The aliens might be hive creatures, colonial organisms like some earthly invertebrates, or even single-celled. The latter is less likely because it is hard to see how organisms with complex capabilities can evolve as single cells. But please note that there is a giant single-celled plant, which looks like some types of algae, that was recently found in wetlands in the forests of the United States–Canada border in the midwest. It is now about 100 miles long and several miles wide, the largest known single-celled organism. It is possible that new types of specialized in-cell organelles, similar to our cells' nuclei and mitochondria, could evolve and provide brain function, for example. Biologists now think that our mitochondria are the result of an ancient bacterial infection at the beginning of cellular life, that proved beneficial to both cell and invader. Such a symbiosis could take other and more complex forms.

In any case, a microscopic and biochemical examination would reveal many differences. These could range from different DNA of our type, to DNA based on other than our four base units, non-DNA genetic molecules, and probably a number of possibilities currently inconceivable. Aliens both biological and mechanical will be taken up in Chapter 9.

Dueling Theories: The Use of Physical Law and Theories to Examine Reports of Actions We Cannot Now Produce; Occam's Razor

We have described above a number of straightforward tests that could be applied in a range of possible cases of the examination of alleged alien artifacts. And we have said that as a first step we would also consider what we observe about them in terms of established physical law. If two alternate explanations of what we observe seem equally correct, we should apply Occam's Razor: choose the simpler one.

"Simpler" does not always mean simple. In some situations it may be a matter of modest degree: if we have no physical evidence to assess, only eyewitness stories, and there is only a complex earthly explanation or there is no clear one, then we are free to examine the possibility of assuming an alien origin. We can see then where that leads.

The first time a walking, talking alien arrives, hands out tissue, blood samples, and food from its planet, displays its pets, demonstrates its faster-than-light ship and its power source, speaks to the United Nations and visits local schools, the *simple* explanation will be, regardless of our abilities to comprehend non-terrestrial technology and biology, that the aliens have landed.

2

GIVE ME A LEVER LONG ENOUGH AND I WILL MOVE THE EARTH

HOW ALIEN SHIPS MIGHT FULFILL ARCHIMEDES' ASSERTION; DEPARTURES AND ARRIVALS; INERTIA

The vast majority of UFO reports are either visual or radar phenomena. In many reports, especially night and radar sightings, a light or group of lights or blips fly in formation. Very often they abruptly change speed or direction. This was also true of Kenneth Arnold's original daylight sighting in 1947, after which he described the flight of the disks he saw flying in formation as if they were stones skipping over water. He remarked at the time that they must have been flying at speeds then unattainable by terrestrial aircraft. Yet, he failed to comment, as have most members of the general public who have seen such things since he did, on the implications of the sudden changes in speed and direction. For those with a grounding in physics it is those changes that raise eyebrows and questions. The reason is that those changes imply large, and in some cases enormous, accelerations— accelerations that would stress the ships and their contents beyond all human experience.

To get a personal feeling of what is meant by high acceleration, consider your experience and knowledge of driving cars. A "hot" car can go from zero to 60 mph in six seconds or less. When you put the pedal to the metal and go zooming off, you are pressed back into your seat with a certain amount of force. When your speed reaches its maximum, the pressure on you drops to zero and you sit comfortably, and hopefully alertly, in your seat. The force that presses on you at its maximum is about one half of the force of gravity (½ g) at the Earth's surface. The pressure you feel is noticeable but quite moderate.

Consider the opposite case. You are driving at a modest forty miles per hour and you glance to the side for just a moment. The moving van in front of you stops without warning and before you can look back again you plow into the van. If you and your car came to a full stop in 1/10 of a second then the deceleration was about 20 g's. If you were wearing your seat belt then you probably would have survived. Even if your car had an airbag that worked properly, you would likely have suffered some whiplash to your head and spinal column. Without the belt and bag you would almost certainly have been turned into a bloody corpse with many broken bones. Your car, of course, would have been transformed into crushed scrap metal.

Now consider the forces on maneuvering UFOs. If the UFO is flying at a rate of 600 mph when first sighted (a respectable jet fighter speed just below mach 1, or the speed of sound), and it shoots off at the same speed at a right angle in a tenth of a second, it undergoes an acceleration of 300 g's. For comparison, fighter pilots pull up to 10 g's in some of their maneuvers, which means they turn in a curve that takes about 3 seconds. To survive just this level of g-force, they wear pressure suits that force blood into their heads so they don't black out. Stresses in the airframe of the plane are significant too and have been known to destroy planes at somewhat higher g force levels. Astronauts launching into orbit go from zero to 17,000 mph in about 10 minutes. That averages to about 1.5 g, though the stress actually rises towards the end as the fuel is used up, reaching 5 g's for a bit. (The record for a human traveling at high g force is apparently 17 g's for 4 minutes, probably set in a giant training centrifuge.)

Another striking aspect of UFO maneuvers is that in almost every case they are completely silent. There is no scream of rocket or jet engines, nor any wind noise from its fast passage. There is no sonic boom, even if the UFO is traveling at an apparent speed estimated to be above the speed of sound.

Now consider the portrayals of space maneuvers in popular science fic-

A mysterious disk photographed over Paris in 1953 seems to be in two places at once, or to be moving so quickly that a ghost image of where it was can still be seen. Photo: copyright 2000 Fortean Picture Library

tion movies and television shows. Leave aside the use of warp drives for a moment and just consider "impulse drives," i.e., rockets. The usual event takes the ship from orbital speed up to some noticeable fraction of the speed of light in just minutes. (If it were hours, the show might be rather boring, or the action would have to cut out and back in with a time jump in between to fulfill the dramatic demands and time limits.) A ship in orbit accelerating from 20,000 mph up to a tenth of light speed (18,600 miles per second) over the course of an hour would feel a force of 1,000 g's. Cut the interval to the dramatically possible 30 seconds and the force rises to 120,000 g's. The ship would rip apart just after every living thing inside was smashed to wet spots in their seats and on the walls. To cut the acceleration to a bearable 1 g would require the ship to ease up to speed over a period of 1,000 hours or so. To achieve that comfortable rate of acceleration going up to two-tenths the speed of light would take about twice as long.

Is there any known technology that would lift the human-tolerance limit on acceleration? Not really. The pressure suits mentioned above help up to the already mentioned 10 g momentary limit. They have a relatively minor effect on your internal state, just preventing the blood from pooling in your extremities. The design of crash couches is focused on the durability of materials against stress. But conventional design ignores some other possibilities, mentioned in the science fiction literature, which are actually today's technology but are rejected because they would bring major weight penalties and force redesign of other systems. These all focus on distributing the g force over every surface as uniformly as possible.

A first conceptual step is to substitute a crash web for the conventional seat belt and shoulder straps in fighters (note: these are entirely omitted in *Star Trek,* and missing also from abduction accounts of spaceship interiors). The web has more points of contact and spreads the load of a fast deceleration more widely over the body. Even so, in a 300-g crash the web would slice an astronaut's body to ribbons. The next step is to provide a complete mesh covering. Then a continuous surface (one must make provision for breathing).

Even better, and this is mentioned in various science fiction stories, would be to flood the ship with a fluid, either water or an inert substitute. This would guarantee, if it filled all the cracks and crannies, that every bit of the ship accelerated in unison. Fluids are pretty much incompressible and this would help rigidify the ship and spread all stresses as evenly as possible—the astronaut's eyeballs would still remain in her head during a very sudden stop.

Even in the case of a flooded spaceship, there is still a problem beyond ordinary materials failure at really high accelerations. People and animals have internal body cavities (torso and skull) that are only partly filled with internal organs. In the flooded ship, the organs would still be free to crash into spinal cord, braincase, or ribcage as the case may be, even though our bodies would be at rest in the ship. We would die looking perfect on the outside, reduced to jelly on the inside. To avoid this fate, the obvious thing to do is to fill our insides with the same fluid used outside. Experiments are under way to demonstrate that deep-sea divers can survive extreme depths by filling their lungs with an oxygenated fluid hydrocarbon. Such experiments have succeeded with rats. The step of filling every cavity has not been tried but probably is feasible, too.

What is the reason these universal problems with acceleration exist in the first place? They are due to a fundamental property of matter called inertia. This property has been recognized since ancient times, but Galileo was the first to describe it clearly and to give it form. Galileo described inertia as the property of matter that tends to keep it in its current state of motion. If at rest, matter tends to stay at rest, and if moving, it tends to continue moving at the same speed and direction. The change from still to moving, or from one speed and direction of motion to another, can only take place if the matter is acted upon by a force. A change of motion is the hallmark of a force in action. The size of the change in motion, the acceleration, is just proportional to the force involved.

When several objects are loosely linked, for example your body sitting on the seat of the car, and the leading edge of the assembly (the car's bumper) hits an obstacle, it slows down first. Then the whole frame slows down, because since it is solid, all parts are in essentially perfect contact. Your body continues on, due to its inertia, at its original speed and direction. Eventually it slows down too, either by hitting the seat belt and air bag or by striking the windshield and dashboard. The car itself crumples because the internally transmitted forces and decelerations are so great that they overcome the internal resistance to stress of the metals and plastics of which the car is made. All materials have limits to their resistance to internal stress.

In spite of Galileo's description of it almost 400 years ago, the true nature and cause of inertia raised questions for Einstein when he was formulating his general theory of relativity, and it is still not completely understood. Einstein noted and was puzzled by the fact that the inertial mass of matter was always equal to its gravitational mass. Inertial mass is what is measured if we use a known force to accelerate a piece of matter a measured amount in a case where gravity is not a factor. For example, consider giving a measured shove to a skater on an ice skating rink, or on a perfectly frictionless surface. Because the shove is horizontal, gravity acts equally throughout and does not affect the acceleration of the skater. The mass of the skater determined from the known force and measured acceleration is his inertial mass.

Now have the same skater stand on a scale or hang from a spring scale or stand on one arm of a balance. Using the measured weight and dividing by the acceleration of gravity at the place on the Earth where the weighing took place (Galileo measured that number, what we have called "1 g," as being about 32 ft. per sec. per sec. at sea level), the gravitational mass of the skater is calculated. These two measured masses, inertial and gravitational, always equal each other. Einstein asked why that should be and did not really come up with an answer. He observed that gravity is a force associated with all matter, but theory then (and still today) did not allow him to predict from first principles the size of the gravitational force between any two well defined pieces of matter, such as between two neutrons. There is an attraction from each piece of matter on every other, as Newton so clearly stated. Of course the strength of it is known experimentally, but no theory predicts it or its strength relative to other forces—electric forces, for example. It has been assumed to be a fundamental property of matter without real explanation. Only recently, with the advent of String Theory, has it begun to look like this property of all matter may someday be predicted from theory.

Inertia is an even stranger case. It obeys no known force law. It only affects each piece of matter individually and seems to provoke no interactions between locally separate masses. The only suggestion for what causes it that Einstein took seriously was made around 1900 by German physicist Ernst Mach. He suggested that inertia is the result of the gravity from all matter in the universe acting on each piece of matter. He proposed that the only correct reference frame in which to measure changes of motion, and thus inertial mass, is a frame made up of the distant stars. It is well known that a gyroscope (absent friction and other effects) points in a constant direction with respect to the distant stars, no matter how it is carted around. This fact has been used to make navigation devices for many years.

Einstein recast this idea in light of his insight that gravity is the manifestation of the curvature and distortion of space-time. Matter causes this distortion. Gravity is transported by force particles called gravitons, which, like the photons that carry the electromagnetic force, travel at the speed of light. The local distortion that acts to cause local inertia in a local mass is the sum of the distortions from all the near and distant matter in the universe.

For matter near the Earth, a super-stable gyroscope in Earth orbit, for example, it is possible to estimate that the inertial influence of Earth, though nearby, is a billionth that of all the distant matter. Even though matter is lumpily distributed throughout the universe and always in motion, always being accelerated by interactions with other matter, on the scale of the universe these irregularities smooth out enough to make it appear that each electron (for example) has the same inertia as any other, no matter where we observe it.

Because the matter in the universe, including the Earth, is in motion, the local distortion of space-time is minutely changing, and in principle this affects the amount of inertial mass for a given amount of gravitational mass. This local distortion could in principle be measured by a gyro in a balloon aloft over the north pole. Suppose the gyro's axis is set to point at the north star. The rotation of the Earth will distort the local inertial frame and try to superimpose a 24-hour wobble on the steady pointing of the gyro axis at its north star. The Earth's changing distortion is only a billionth of the steady inertia from distant matter. The effect, therefore, is tiny. But the experiment is in the works.

If we return to the UFO phenomena mentioned above in light of this knowledge of inertia and acceleration, we can make several comments about possible alien technologies. First, if the reports of sudden changes of direction are taken at face value then the UFOs, assuming they carry living passengers, must somehow have total control of inertia. Even if UFOs are only instrumented probes this would also likely have to be true, or they would have to be made of materials unknown to human science, because the internal stresses they undergo would destroy known materials and alloys.

The *Star Trek* solution is to have "inertial dampers," which can turn on and turn off inertia in the ship at will and nearly instantaneously. How might these work? No one knows, but one possibility is to create gravity artificially in the direction opposite to and of strength equal to the force of acceleration, so that they cancel. Or, if zero g is to be avoided, the cancellation would leave a normal 1 g left over so that the astronauts could remain standing normally on the command deck.

The UFOs also apparently control what goes on in the air in their vicinity, too. The reports of high g maneuvering state that the UFOs are usually silent. The majority of UFO encounters, where a craft is alleged to have been seen in some detail, put their sizes in the range of jet fighters. In the cases when consideration of the data suggests that they are traveling at supersonic speeds, no sonic booms have been reported even though they would be a normal accompaniment to any supersonic aircraft flown by humans. The lack of air noise and sonic booms is a real puzzler. Any solid object generates a pressure wave as it moves through a substance, such as air.

That UFOs fly in silence has been a long standing puzzle and led many enthusiasts to suggest that they are propelled by some sort of silent antigravity drive, using the gravity of a planet or star for their lift. But even if such a drive existed, it would not solve the problem of lack of air noise or sonic booms.

WAVES OF GRAVITY

To think about gravity drives and antigravity, we need to inquire briefly into the nature of gravity as it is now understood. Newton thought of it as a natural force that acted between all matter in the universe, always attractive, and with a definite force law or equation. The equation states that the force is directly proportional to the amount of matter and inversely proportional to the distance between the matter squared. The first part is intuitively obvious: the more matter, the stronger the force. The latter means that if two pieces of matter are pulled apart from each other, when their separation is twice the original value, the force of attraction between them is $\frac{1}{4}$ as great, when the separation is three times as great the force is $\frac{1}{9}$th as strong, etc.

Newton considered the question of how pieces of matter can attract each other when not touching, and he simply observed that such was the case, no matter that no theory explained it. He suggested that the force was felt instantly from any bit of matter by any other. In other words, the force traveled at infinite speed. He was not entirely comfortable with this, but he had no data to contradict it. He also suggested that all space surrounding matter can be thought of as being pervaded by the gravitational field of that matter, and that the field can be represented by a map of contour lines whose spacing indicates the strength of the field at each place in space. The gravity field surrounding a sphere of matter is shown as a set of radial lines, which spread out in space as they extend to infinity. Think of the spines of a sea urchin. The field fills space, but, according to Newton, does not interact with it.

This picture works extremely well and served for about 250 years without alternative. It is still used in many terrestrial and astronomical contexts, though improved technology has revealed over the years small discrepancies with nature. These discrepancies, along with logical inconsistencies such as the supposedly infinite speed of gravity, led Einstein to completely rethink gravity. The result was his general theory of relativity. In general relativity, Einstein considers gravity to be an interaction between matter and the space and time it is imbedded in, an inherent and inescapable interaction. In his view, all matter warps space and time, and the degree depends on the amount of matter and its density.

This warpage affects all other matter so that each piece follows the shortest path possible in the warp of its local space. We experience that in terrestrial conditions as normal gravity. Einstein's complex equations for all this reduce down to Newton's single simple equation in "normal" conditions. To see obvious discrepancies between Newton and nature, you must encounter abnormal conditions, such as travel near the surface of a neutron star (a star with the mass of the Sun packed into a sphere the size of the Earth).

Einstein also showed that gravity travels at the speed of light rather than at infinite speed. In his view the force of gravity is carried by gravitons, particles analogous to the photons that transmit light. Also analogous to photons (and tiny particles of matter), gravitons can behave as both particles and waves, depending on the circumstances. In a sense, one can think of the force of gravity not only as space warpage but also as the interchange of gravitons between masses. Gravitons can attract each other as well as attract masses. This behavior is more complex than that of photons, which can interfere with each other but not attract or repel each other.

Gravitons are emitted in great bursts in certain circumstances. These include the destruction of matter, of which the most violent example is the explosion of a supernova. There have been experiments to detect graviton bursts using large aluminum cylinders instrumented with extremely sensitive strain meters to detect the minute vibration we would expect to see from the passage of such a burst of gravitons. These experiments failed. New experiments using much larger and more sensitive laser strain gauges are now under construction to look for gravity waves from supernovae and other sources. It is expected that they will be detected because Einstein's theory (and the even newer string theory) call for them.

There is also one excellent piece of indirect evidence for gravity waves, which won radio astronomers Joe Taylor and Russell Hulse the Nobel Prize for Physics in 1993. They observed a pair of radio pulsars, leftovers of two stars going supernova, in orbit around each other. Each pulsar emits a precisely timed set of pulses that is slowly decreasing in frequency as the neutron stars spin down over thousands of years. Their periods are precisely determined and this allows us to see other effects. Every orbit, the pulse frequency increases as the pulsar comes toward us and decreases as it moves away. This variation occurs once each orbital period, so that period can be precisely determined as well. Taylor and Hulse used the orbits they determined using this data and Kepler's orbital laws (dating from the 1500s) to find the stars' masses. Einstein's theory predicts that for two such large masses in such a close orbit the system will radiate gravity waves, which take their energy from the orbital energy. Thus the orbits will shrink, the stars' orbital speeds will increase and the orbital period will shorten at a calculable rate. Taylor and Hulse observed the period to be decreasing in accord with the predictions of general relativity. The implication is clearly that the system is radiating gravity waves.

THE POSSIBILITY OF ANTIGRAVITY DRIVES; THEIR UTILITY IN SPACE TRAVEL

Now that we know that gravity waves exist out there, can they be yoked to serve space-traveling purposes? This is not yet at all clear. Perhaps there might be a way to use gravity as it exists to cancel out the gravity of a planet. One brute force technique would be to provide a local source of gravity equal in strength to the one you want to cancel, placed so it pulls in the opposite direction. After all, on the line connecting the Earth to the Moon, there is a region where the Earth's and the Moon's gravity approximately or exactly balance. (Because the Moon exerts ⅙th the gravity of Earth, this region is about ⅚ths of the way to the Moon.) Of course, to make this effective you would have to be able to move this canceling mass as your ship moved. Conversely, if the canceling mass were much bigger or closer, it would pull the ship along.

Using wave interference, which works for photons, one might try to cancel the local gravitational attraction using some local controllable source of gravity waves. Its strength would have to be comparable to that of a planet (or star, if the ship was an interstellar one). But we don't really know how to make such a device.

Suppose we did have a gravity nullifier, how could we use it? On Earth it could be used to make moving around heavy objects considerably easier than normal. It would reduce to nearly zero any friction with the Earth's surface. Moving a heavy load would only require the initial push to overcome inertia, and a counter push for the same purpose to stop it. These would not be negligible efforts if the load were massive, but no energy would be needed to keep the load moving. Another use would be to allow buildings to stand essentially weightless. Their structural requirements would be reduced to those needed to withstand wind and earthquakes. This would allow a large savings in construction costs, assuming the nullifier were cheap. Having a personal nullifier

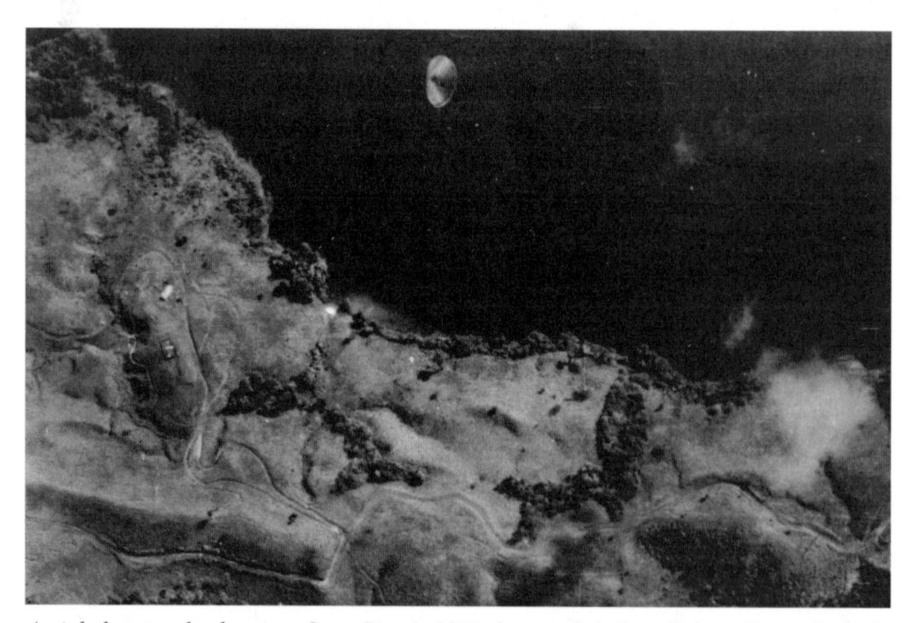

Aerial photograph taken over Costa Rica in 1971 shows a disk-shaped object that seems to be moving while tilted vertically. Anti-gravity? Inertial dampers? Photo: copyright 2000 Fortean Picture Library

A student from Mexico State University shot this object in 1967 while photographing land formations for a geology class. Again, it seems to be a disk flying on its side. Photo: copyright 2000, UPI/Corbis-Bettmann

would give you a new spring in your step. In space, a ship would need only the energy to overcome inertia and escape from the gravity of a planet or star, which would be much less expensive than with gravity "on." This would still only be a gravity nullifier, useful, perhaps, but not antigravity.

How could antigravity be achieved? The most obvious possibility would be to find a form of matter that is antigravitic. Such matter might be repelled by ordinary matter. For a while it was thought that the antimatter we already know about might have that property. But experiments were run in a subatomic particle accelerator in which antiprotons were created during high energy collisions. Their paths in the accelerator were carefully examined to see if they fall upward instead of downward as normal protons do. The result was that they fall downward at the same rate as protons, which was rather disappointing. If such antigravity matter exists, no theory predicts it.

Another possibility is to create space curvature opposite to the curvature associated with normal matter by artificial means. Again, no method for doing that is known. It should be noted that, if such a thing were possible, to have a major effect the antigravity strength would have to overcome that of normal gravity. This would probably require an artificial mass or its functional equivalent with a gravity greater than the Earth or the Sun. This seems on its face to require so much energy as to be out of the question.

Suppose, however, antigravity exists. How would it work? A spaceship equipped with antigravity matter could simply drift away from Earth or the Sun at a rate depending on the amount of repulsion. That, in turn, would depend on the amount of this weird matter and its separation from its departure point. Presumably Newton's law of gravity would work in reverse and the repulsion would weaken with the distance squared, just as the normal attraction does. It would not provide very rapid acceleration.

There is another problem. Steady natural antigravity would allow easy departures but arrivals would have to be powered to force the ship to its destination against steady repulsion. Once there, it would have to be tethered to the ground somehow, otherwise the antigravity repulsion would make it drift away. An antigravity ship would float up off its spaceport pad.

To achieve rapid acceleration and avoid landing and mooring problems the antigravity would need to be artificial and controllable. Then one could turn it on and zoom off, shut it down gradually to land, and shut it off for docking or landing on the ground.

If one had an artificial antigravity drive, it likely would be silent and invisible, but rapid motion through the air would still incur air noises and sonic booms, which as already stated are not observed of UFOs. There would probably also be more exotic and disturbing phenomena. The extremely small region of artificial gravity might warp space only locally. This might create a region of great optical distortion around the ship, because light bends to follow the space warp. In space, when the artificial gravity came on, the stars would appear to shift, and if the ship were near Earth and you looked through the warp region at the Moon, it would look fattened and distorted.

WOULD ANTIGRAVITY-DRIVEN SHIPS DESTROY WHAT THEY VISIT? THE EFFECT OF AN *INDEPENDENCE DAY*-STYLE MOTHER SHIP ON THE EARTH AND THE MOON

If employed in taking off from or maneuvering near the ground, the local distortion would likely cause soil to fountain and houses to fly apart. A person standing nearby could be severely injured. The effect would be that of an extreme artificial tide. (The tides are simply the result of the strength of gravity from a piece of matter varying across space, because it gets weaker with distance.) If the source of gravity is massive and extremely compact, such as a neutron star, and you fly near its surface with your feet down and head up, just that small difference in distance from the star will create vastly stronger gravity at your feet than at your head. The difference will overcome the molecular binding force of your tissues and you will be ripped apart.

This problem would force the antigravity designers to create a field that was zero or 1 g inside the ship while strong outside to avoid the crew undergoing "rip tides." It is not clear that such a thing is possible.

An *Independence Day*-sized mother ship, essentially a flying city one mile across and powered by antigravity, would be so large that it would have gross effects as it arrived on Earth that are not portrayed in the movie. If it came screaming in from outer space at "just" thousands of miles per hour (an interplanetary but not interstellar speed), it would create a shockwave that would pulverize any city below it. No death ray, as shown in the movie, would be needed. If it slowed down outside the atmosphere and came in on antigravity drive, the field required to repel it might still be so disruptive to Earth as to create a giant soil fountain. It seems the general rule, no matter the type of drive, must be: that there can be no big accelerations near any planet or inhabited place. Otherwise, the point of departure or destination will be damaged.

WOULD ANTIGRAVITY KILL INERTIA? PROVIDE HIGH-G SAFETY FOR PASSENGERS?

It might be asked whether the "simple" achievement of an antigravity drive would also solve all the problems with inertia and acceleration. It is not obvious that this would be so. We pointed out that the inertial and gravitational masses always appear to be equal to each other. But the former arises from the local sum of gravity from the universal distribution of matter, whereas the latter is mostly the result of the size and distribution of local matter. Canceling the local effects of local matter will not necessarily cancel the local effects of distant matter. It seems likely that separate inertial damping will be needed to deal with extreme accelerations, even if antigravity drives can be utilized. In a later chapter, we explore the use of local space warps as faster-than-light drives. In that case, with the ship completely enclosed in a local warp, internal inertia would probably disappear, and this would relieve the ship and crew of the need for extreme anti-inertia measures, such as flooding the ship and every living thing in it.

3

TRAVELS WITH ISAAC AND ALBERT

CLASSICAL AND RELATIVISTIC VISITS FROM THE NEAREST STARS

The numerous accounts of UFOs maneuvering in the atmosphere of Earth and accounts of abductions by aliens beg the following questions: If it is the case that aliens are arriving and departing all the time, where are they coming from and where are they going to? And what means of travel are they using? We have just a few clues based on the published accounts. For the most part, the aliens are not revealing their origins or their itineraries. To make progress on these questions and on their modes of transport, let us ask what observations show us.

First, the great majority opinion has it that these are visitors to Earth we are dealing with, not hidden natives. A few writers have alleged that they are dwellers in secret subterranean locations, but this view is held by a tiny minority. No one ever seems to see the UFOs emerge from inside the Earth, though occasionally they have been alleged to surface from under the ocean. But the statistics of sightings greatly favor atmospheric phenomena and landings from above.

Let us assume, as most believers do, that they are visitors from outer space. Then, it would appear there are several possibilities: they might have bases on the Moon, one of the planets or their moons, or the asteroids, or on a gigantic mothership somewhere in our solar system. Otherwise they must come directly from a more or less distant star system.

Suppose they are based in our solar system. We have good reason to think that if they breathe our air and drink our water, as most accounts clearly suggest, then no place off Earth in our system can be their place of origin. The other planets and moons are too hostile and too different for that. On the other hand, they could have bases that encapsulate an environment friendly to them, as we have done in our space stations, our landings on the Moon, and will do in our proposed visits to Mars and the moons of Jupiter.

If they are traveling regularly around the solar system, would we see them if they are out in space or maneuvering near a large ship or a base on any other heavenly body? The answer to this question is: "that depends!" Let us assume a "Standard" Alien Landing Craft that holds 20 aliens and their supplies for trips of up to two weeks. This allows transits from Earth to the Moon or to a large mothership in distant Earth orbit. I use this passenger load based on the somewhat vague accounts of abductions to alien ships. The number of aliens directly involved with the abductee is often in the range of 5 to 10. It is not unreasonable to assume they are the medical/biological crew, and that there is an equally large crew devoted to astrogation, piloting, maintenance, cargo and command. A ship with technology that provides for crew needs not too hugely advanced beyond ours might be scaled up from the parameters of the space shuttle. A crew of 20 is about three times the shuttle's and two-week orbital missions are about the same mission length as the shuttle, so the alien craft might mass about 3 times the shuttle and this could require the craft to be about 1.5 times the shuttle's outside dimensions. The engines would need to have more than three times the shuttle's thrust.

Remember these dimensions were developed assuming that the alien ship would operate like the shuttle in Earth orbit. It could maneuver in orbit and make one trip down to the surface. Going back to orbit it would need to somehow acquire the equivalent of the shuttle's booster. (The shuttle has a booster with two additional main engines and a lot more fuel than on the orbiter.) Otherwise it must land with its booster-equivalent attached, a nasty design problem with no simple solution that maintains the assumed small spacecraft size. If, as many UFO accounts suggest, it were to be able to maneuver extensively into and out of the atmosphere, and land and take off repeatedly, the energy requirements would rise dramatically and no account of any landing or close-up sightings matches the scale that would be needed for a chemical rocket of current human design.

Also assume, to begin with, that the propulsion is by chemical (our kind) or nuclear (only lab tested so far) rocket, which would create the most visible disturbance for a rocket system that allows quick maneuvering at conventional speeds. We can then ask, Would a rocket flame with three times the shuttle's power (or any other feature of the craft) be visible as it is ignited in space near Earth? The answer is that the plume would be unlikely to be visible to the naked eye. While smoke and flame from the shuttle's engines are overwhelming at launch, the flame is nearly transparent in space. It has rarely been videoed either from space or from the ground. For comparison, look at the videos taken in the later Apollo missions by cameras left on the lunar surface when the LEM took off. There is just a bit of dust and abandoned garbage kicked up, but no flame except exactly at initiation of ignition. The atmospheric conditions were the same as in space—no air!—though it is true the thrust of the LEM was smaller than that of the shuttle, which would make a slight difference. But it looks like rocket plumes in space are difficult to spot, and it may be that a chemical rocket with many times the shuttle's thrust will be invisible from a distance of as little as hundreds of miles if viewed sideways by naked eye. Perhaps if seen nozzle-on the brightness would show up as a pinpoint of light. A nuclear rocket, which might run much hotter, would likely have even less of a chemical plume, but might be visible in the ultraviolet.

All these remarks apply to naked-eye observations, but also reasonably well to telescopic ones too, as far as seeing the flame. However, seeing the craft itself is perhaps another matter. There have been video images made by amateurs on the ground of the shuttle, using TV cameras attached to modest telescopes. These showed the shape of the craft and its open bays. But the shuttle was only 120 miles up. On the other hand, we should not ignore more advanced techniques and equipment. The Hubble Space Telescope could capture an image of a Standard Alien Craft—at least as a bright point—if it had a visible-light-reflecting surface and was as far away as the Moon.

Several space-tracking operations have similar capabilities. Brian Marsden, director of the International Astronomical Union's astronomical event notification system, located at the Smithsonian Astrophysical Observatory, recently mentioned two incidents to me that shed an interesting comparative light. The first is that of the object 1991vg, optically observed at a distance of 0.03 AU (1 AU—or Astronomical Unit—is the radius of Earth's orbit, about 90 million miles), which is about 2.7 million miles away, or 10 times the distance to the Moon. It's orbit was almost identical to Earth's, and it is unsettled whether it is natural or artificial. If the latter, it is not clear that it is of terrestrial origin. The size and mass are still not known and it will be more than a decade before it falls out of alignment with the Sun so that it can be seen again. When last observed, it varied slightly in brightness with a period of less than a minute, suggesting an aspherical shape tumbling in space, perhaps a rocket booster, or perhaps, Marsden speculated, a large "chip" off the Moon's surface, the result of an ancient impact.

Second, during the Leonid meteor shower last fall, visual observers reported seeing brief flashes of light on the Moon during the hours when the shower was most intense on Earth. Flashes have been reported sporadically at non-shower times over the years but have not been confirmed. In this case, more than one observer confirmed the flashes, and even better two or three independent videos showed flashes at the reported times. It seems likely that the flashes were real and that the Leonids caused them. There is a puzzle, however. The Leonids are thought to be only dust size. If true, even at their considerable relative velocities up near 35,000 miles per hour their kinetic energies would not produce a flash at impact visible from Earth. It is possible that they kicked dust up from the lunar dark, near the terminator where they struck, and into the sunlight above, where it reflected strongly to us against the dark lunar night.

Marsden also said that the optical patrol programs looking for potential Earth-colliding asteroids have detected objects as small as tens of yards as they pass as close to Earth as the Moon. Furthermore, the US Air Force (and presumably the Russians) maintains an optical space watch that tracks everything in near-Earth orbit larger than a basketball.

The great powers also maintain radar early-warning and space-tracking systems that are capable of detecting objects of similar size and distance. All these observations suggest that ships themselves—if not black in visible wavelengths of light, if solid, if not somehow radar-stealthed—would be likely to be detected as they operate near Earth and in space out to the Moon's distance, at least.

We can also ask, if there is a lot of alien ship traffic in our vicinity, how do they communicate with each other? If they do it by radio, we should have heard them. There is a small but growing group of radio astronomers with supersensitive equipment observing the sky almost all the time; another group looking specifically for intelligent signals from space; the various intelligence agencies; and thousands of sophisticated amateur radio operators all listening hard. Because we can hear lightning storms on Jupiter and also the 4-watt whisper of the Voyager craft out beyond Pluto, the alien signals should be obvious. Because no one has yet picked them up, it is reasonable to conclude that the aliens, if here, are either keeping radio silence or communicating in a band or spectrum we do not receive, or somehow hiding their signals in plain sight.

GETTING THERE IS 99.999999 PERCENT OF THE FUN

We concluded above that the aliens might have bases in our solar system, but originate in another star system. They must get here from there somehow, and many methods leap to mind. They all face one overwhelming problem, however: the immense gulf between the stars. In the Sun's neighborhood, the nearest star is 4 1/2 light years away, and the average distance between the stars is about 10 light years. If the fastest thing we have ever launched, the Voyager II spacecraft, which is traveling like a meteor at about 35,000 miles per hour, were aimed at the nearest star (it wasn't), it would arrive in about 84,000 years.

To travel to the nearest stars in "just" a few decades would require interstellar ships that can travel at significant fractions of the speed of light. If ships can achieve speeds above about 1/3 the speed of light then effects predicted by Einstein's theory of relativity will become important for the ship's performance, for the lives of the astronauts on board and for what they observe of the universe through their ship's portholes (if any) and viewing systems. Is there a way to achieve near-light-speed performance?

PRIME MOVERS: ROCKETS, FISSION AND FUSION RAM ROCKETS, AND ANTIMATTER DRIVES

According to current theory, it will be possible to improve on our chemical rockets by switching fuels, and in doing so we might well be able to increase the speeds we can attain by a factor of two or more. This would require the development of some new materials to withstand the higher nozzle temperatures. Nuclear fission rockets have the potential to improve things further. The improvement would be well worth looking at for planetary travel, and the United States has carried out preliminary experiments on atomic rockets. However, the improvement would not be more than another factor of 10, and that doesn't carry the day for interstellar travel. There would be potentially significant issues of radioactive contamination for us, and presumably for ethical aliens. (Incidentally, though radioactive contamination has been claimed at a few UFO landing sites, the amounts are too low for the effect of a nuclear rocket takeoff, and the composition of the radioactive elements has not been shown to be compatible with any known or speculative power chain reaction.)

If chemical and fission rockets are out, then the search for locomotion must turn elsewhere. There are two "conventional" possibilities in the sense of more or less known physics: the fusion drive in its various forms and the photon sail drive.

Concept designs have been published for rockets that fuse hydrogen nuclei into helium. They are simple in principle, generally consisting of a magnetic bottle capable of compressing the hydrogen gas up to temperatures and densities typical of the core of the Sun—that is, tens of millions of degrees and hundreds of times the density of water. But in practice these are difficult to achieve. The worldwide effort to make fusion reactors based on the magnetic bottle and other techniques has been going for 40 years. It has cost billions of dollars and success is still in doubt. In any case, it is at least a decade away.

The power available from fusion is much larger than for any of the previously described schemes, but, even so, the energy needed to travel from star to star at high speeds is huge. To get a mental picture of just how big, consider the fictional starship *Enterprise*. The fusion fuel needed to propel it up to say half the speed of light, and also allow for deceleration and a return trip, would actually take up a volume about 400 times that of the ship, which is portrayed as more than half living and cargo space. You do not see that fuel tank. This gigantic fuel supply would allow just one round trip which, to the nearest star, would take 16 years of transit time, back and forth. In 1968, Freeman Dyson estimated that a particular "best case" version of a fusion rocket, one which used fusion pulses from exploding H-bombs, might be feasible. It could accelerate the ship at 1 g for ten days, achieving a coasting velocity of 3⁄100 the speed of light (the one-way trip to the nearest star would be 120 years). The four-million-ton ship would have to explode 300,000 bombs and would cost about 1⁄10 of the whole U.S. GNP for 1968.

The British Interplanetary Society did a study called Project Daedalus, finished in 1978, which examined the possibility of using fusion pellet microexplosions instead of bombs. The requirements on the strength of the rocket structure are relaxed and the burden of fuel decreases. It would accelerate continuously for four years to get up to 12 percent of the speed of light. For a one way, non-stop voyage to the nearest star, the trip would take about 35 years. A round trip would take close to 120 years (the extra time is required for slowing down). The payload of 450 tons (people and their ecological support systems and instruments) would require about 55,000 tons of fuel for the one-way, nonstop trip.

If aliens somehow have finessed the problem of extreme acceleration that would be required to jump from zero to half the speed of light in a few hours, time (instead of over a year), the power draw would be almost a billion times the total power now consumed by all human activities in the world combined.

Where would one find huge quantities of hydrogen? Earth's oceans have plenty tied up in water, but the penalty of boosting it into orbit is too high a price to pay. It is possible that a comet could be captured or that a ship could scoop methane out of Jupiter's or Saturn's atmosphere. The fuel transportation energy for those possibilities is within the realm of possibility. But all this fuel would have to be carried on board for the trip, and that is a huge penalty on a given payload.

By the way, the recent discovery of extrasolar planets (the count is 40 as I write in spring of 2000) provides an additional interesting possibility. These planets are all roughly Jupiter-sized and presumably also gas giants with roughly the same composition as Jupiter. They are being found so frequently that it is likely they would be found in almost every stellar planetary system, and thus interstellar travelers could use them for refueling.

Understanding the great penalty one pays for on-board fuel, and knowing that hydrogen is by far the most abundant and widely distributed constituent of the universe, it has been suggested that, except perhaps for a small reserve, hydrogen fuel could be scavenged from the gas clouds between the stars along the journey's route. (Robert Bussard published a discussion of the idea in 1960.) This leads to the notion of a magnetic bottle engine with a huge front-facing open magnetic or electric field scoop, many miles across, which feeds the interstellar hydrogen into the fusion chamber and then into a magnetic exhaust nozzle. The logic of the design problem suggests that such torch ships, or fusion ram jets as they are often called, will be elongated in design, with crew quarters in front to keep them safe from the high energy radiation emitted by the fusion reactions. If one thinks of designs of UFOs, saucers would be out but cigar-shaped ships might be in.

Carl Sagan considered a 1,000 ton starship that would accelerate continuously at 1 g using interstellar hydrogen as fuel. He assumed a magnetic scoop diameter of about 2,800 miles, which would suffice if it could feed off a density of hydrogen of 1 atom per cubic centimeter, a fairly typical average density in the solar neighborhood (but an extremely high vacuum on Earth). The density varies from much lower numbers in deep interstellar space between our galaxy's spiral arms, to tens of thousands of times greater in giant interstellar clouds where star formation is taking place. With continuous acceleration, the ship would approach the speed of light in about a year and the round trip travel time to the nearest star would be about 10 years.

The magnetic field scoops have to be generated somehow. To keep their energy requirements to a minimum it is assumed that they will be generated by superconducting magnets, because these yield huge field strengths and do not require much energy for their maintenance. The technology for these is rapidly developing, and getting the required field strengths will probably not be a problem. However, the actual magnets will sustain very large stresses from the field as the pressure on the field grows, due to the impact and steering forces exerted on it by the on-rushing interstellar gas. (Consider an umbrella in a hurricane.) These stresses have been estimated to be so large that it might not be possible to have a magnetic scoop with a diameter larger than about 1,400 miles. Thus, the performance of any ramscoop ship may be limited to accelerations well under 1 g.

Could we see one of these fusion ram-jets if it were in our night sky? The answer is yes. The fusion flame would be bright in the visible and ultra-violet due to its extreme temperature and the particular emission spectrum of hydrogen and helium. It would also give off gamma rays and X rays, which could be seen with suitable detectors. Seen side-on the jet might look like a tiny needle with an intensity like a welder's arc. So far, no UFO eyewitnesses have reported this. Such a propulsion system would be suitable for space but not for planetary landings and take-offs.

The annihilation of matter by antimatter would be an even more powerful source of energy than fusion. Just what is antimatter? For every ordinary (read terrestrial) type of subatomic particle we know of there exists a counterpart, which is called its antiparticle. For example, the proton has its antiproton, the electron its antielectron (usually called a positron), the neutrino its antineutrino. The charged antiparticles have the opposite electric charges to our "normal" ones. The positron is, as its name suggests, positively charged, while the ordinary electron is negatively charged. (Keep in mind that the labels "ordinary" and "anti-" are as arbitrary as positive and negative, and simply reflect the order of discoveries in this field.) One might ask if the antiparticles enjoy antigravity. This was considered a real possibility until experimenters carefully observed the rate of fall of antiprotons in a subatomic particle accelerator; it was normal. We create the antimatter for these experiments by causing certain subnuclear collisions to take place. In nature, matter-antimatter particle pairs are created out of high-energy cosmic-gamma-ray decays. We see a small amount of natural antimatter in the cosmic ray showers that continuously bombard the Earth.

Though both kinds of matter should have been created in equal amounts at the moment of the Big Bang, our kind of matter predominates here and as far out into the universe as our detectors can see. This appears to be the result of a slight predominance of matter after the earliest instants of the universe's existence, and it might be due to a subtle asymmetry in the rates of certain subnuclear reactions.

For this discussion, the salient fact is that when matter and antimatter meet they annihilate each other completely. By comparison, fusion and fission nuclear reactions convert only a few thousandths to a few hundredths of the involved mass into energy. Thus a little annihilation will, by comparison, go a long way! The fuel supply for onboard storage would drop from the factor of 400 times the ship's mass to around ten times. It's a major improvement, but remember this is fuel for just *one* round trip!

You might also inquire as to whether natural interstellar antiprotons could be scooped up along the way with the protons. The answer is yes, but two things stand in the way of making this an effective drive solution. The first is that since antiprotons have a charge opposite to protons, the magnetic scoop will steer them in the sense opposite to that for protons. It may be possible to deal with this. More difficult is that antimatter is extremely rare in nature, as alluded to above. There is not enough of it to annihilate all the available normal protons at the required rate with scoops sized for normal hydrogen alone. Going to larger scoops is probably physically impossible due to the stresses involved, and thus the efficiency advantage of annihilation is wiped out for the ramscoop.

TO SAIL THE DEEPEST OCEAN: PHOTON SAILS

The advantage of not carrying fuel on board is so great that every propulsion system that successfully avoids this problem is worthy of consideration. Another leading category of alternative propulsion systems are those that get their motive power from the pressure of light. Every photon of light exerts a tiny bit of pressure, proportional to its energy, on whatever it strikes. Even in bright sunlight, the pressure per square inch is minuscule, but in the vacuum of space it is not negligible. It will drive a well designed sail to significant fractions of the speed of light in appropriate circumstances. Of all interstellar drive proposals, this one appears to require the least new technology.

You might think that the ideal photon sail would be black, a perfect absorber, because then all the photons that strike it would be absorbed. We can already make nearly perfect black surfaces. But it turns out we can do better by making the sail into a mirror. Then the photons bounce off, and in doing so they transmit to the sail twice the momentum that would have been deposited on the sail if they had been absorbed. The reason is that the reflection converts the photon's "positive" momentum into its opposite. Call the original momentum a +1. The reflected photon has a momentum of -1, and the change in momentum is +1 -(-1)= +2. So what we want is the perfect mirror. Here again modern technology, developed chiefly to make efficient lasers, can supply our needs. Commercial mirror coatings are available that reflect more than 99.99 percent of light in all visible and some infrared wavelengths.

These coatings are all made on some sort of support, such as metal, glass or plastic, including the now wide-spread thin-film aluminized mylar plastic films. These would work as solar sails now, except for one major problem. They are all, including the plastic film, too dense to make an efficient photon sail. The denser the sail the greater the penalty on payload (or the greater the size of the sail for a given payload). Ideally, the sail should be a gossamer mirror, made of a single layer of atoms. This seems unlikely to be realized because the most efficient mirror coatings are usually about 20 to 50 layers, each several atoms thick, and these are not self supporting. Even so, if a substrate could be found that is quite strong and only 100 atoms thick, it would be a great improvement over the thinnest available aluminized mylar.

Additional weight savings might be accomplished by perforating the sail. The sail might be made with tiny pinholes, perhaps 1,000 atoms across, spaced apart at random by a comparable or smaller distance. This would save about 50 percent or more in mass, and the sail's acceleration potential would rise proportionately. The sail would still act as if it were solid for light, from the near ultraviolet throughout the visible and into the infrared, because the corresponding light waves have wavelengths four or more times longer than the openings—for them, the sail acts as a solid continuous mirror.

Where is the best place to launch a solar-driven photon sail? Because you need as much light intensity as possible, the best location is as close as possible to the Sun. At one solar radius above the Sun's surface a sail-ship would be 100 times closer to the Sun than if it were in Earth's orbit and the intensity of sunlight would be 10,000 times greater. This would be an excellent launching location, but working so close to the Sun would require shielding the ship and its sail by a shadow caster to hide it from the blazing light until the moment of launch. At the moment of launch, the occulting object would move aside or the sail would be unfurled. The ship normally would be dragged along behind the sail on the sunlit side. It would, therefore continue to need protection from overheating. A highly reflective sail of the sort desirable to provide maximum efficiency would also absorb such a tiny fraction of the incoming light that its temperature would remain under 1500° F, and thus a number of sail materials already in hand could survive. Perhaps the ship would deploy some sail material over itself for protection at launch.

The sail would be so thin that to maintain it as a flat surface it would likely have to be spun. Alternatively, it could be designed to have a parachute shape, but would still have to be spun to hold its shape.

The achievable and usable acceleration will depend on both the lightness of the sail and on the type of payload. The hypothetical range of sail materials might allow from 1 up to 500 g's of acceleration. If just an instrumented probe, the high end would be desirable, if it could be achieved. A boost phase at 200 g's that ran for 10 hours would top out at 2⁄10 of the speed of light. This is a very respectable speed, one at which relativistic effects would not be important but could still reach the nearest star in about 25 years. If the cargo includes humans then the acceleration must be limited to about 1 g, because higher g's will eventually do us damage (of course we can't speak with certainty about alien bodies) in ordinary living conditions. With a crash couch, the highest acceleration humans have pulled without damage or blackout is about 17 g's for four minutes. One g would be a good thing in any case to avoid the effects of muscle tone degradation and bone loss seen in astronauts in long-time Earth orbit at zero g. A 1 g boost would also avoid special packaging of all the other living things on board. If 1 g could be sustained for about three months, the manned probe could also reach 2⁄10 the speed of light, and get to Alpha Centauri, the nearest star in 25 years. During the coasting phase, the bulk of the trip, the ship itself would need to be spun to provide an artificial 1 g.

LASER SAIL LAUNCHERS AND SOLAR CONCENTRATORS

Ordinary sunlight is not the only light available to us. We now have powerful lasers that in principle could remain tightly focused over interstellar distances and thus provide a source of light pressure that hardly fades at all. We could also make a huge solar concentrator out of sail material and intensify the sunlight on a smaller light sail, which would enhance acceleration.

The calculations done by physicist Robert Forward and others suggest that one could deploy a solar-powered super laser in the inner solar system, where solar energy is intense and the solar cell collecting area could be minimized. At the orbit of Mercury a solar cell collector with the current conversion efficiency of 10 percent would need to be 600 miles in diameter to drive a manned mission. The lasers would put out 10^{16} watts, about 1,000 times the total current world's output. The laser light would be directed by a large telescope to keep it focused over distances of light years. To stay focused out to Alpha Centauri would require a telescope of the same size as the sail, about 600 miles in diameter. This size setup could transport about 3,000 tons out of the solar system and boost it to about $^2/_{10}$ the speed of light, again making travel time to the nearest star about 25 years. The sail stress issue remains important, but with the higher intensities achievable with lasers the sail may be made of heavier and stronger materials, thus avoiding the limitations of the thinner sails.

To make a round trip using a solar/laser light source, the sail would have to be made of several pieces, some of which would be left at the destination star. This causes a weight penalty, though perhaps the sails could be maneuvered into orbits at the destination where they could be picked up and used by later visitors. The most elegant scheme at the moment is for the sail to be made of nested annular rings. The outer ring would be detached, pushed ahead of the rest, and reshaped to focus light backwards onto the inner 220-mile-diameter sail. The inner part would slow down and go into orbit around the destination star. For the return, the outer piece of this would detach and focus the beamed light on an inner 70-mile-diameter disk. This would return the ship to Earth, where it would be slowed by the laser or solar concentrator.

A second possibility would be to trade sail area for a second laser system, which would be transported and installed in the destination star system. Then the crew would have the choice of returning or continuing on, perhaps having used local materials to resupply and to construct another laser system to carry to the next star.

The sail shape described, of nested rings, is not only efficient as a sail but with appropriate design can also serve as the huge concentrator of laser or solar light. The modification is to specify the diameters of the rings so that they satisfy a particular equation devised by the French physicist Fresnel in the early 1800s. The lens version of this design, called the Fresnel lens, is in common use as the lens for lighthouse lights, for theater spot lights, and as those plastic lenses often seen on van rear windows, which give a reduced-scale wide view of the road behind. Some authors have envisioned making the concentrator as a lens. This works well for the one-color light of a laser but causes focusing problems (chromatic aberration) for the broad spectrum of sunlight. A mirror version avoids this problem and would weigh less.

If aliens had inertial dampers to suppress all acceleration stresses on the ship, sails and shrouds, and perhaps tractor beams to deploy and hold the shrouds, then the efficiency of the sail would be greater, acceleration limits would be removed and photon driven ships might approach the speed of light. Then trips to the nearest stars would take "only" decades as time would be counted by those left behind.

In any case, there are no reports of UFOs that suggest any sort of photon sail ships have visited us here. A light sail 200 miles in diameter traveling near Earth and face on to us would be easily visible. In orbit 24,000 miles up (a so called "stationary orbit"), it would be a mirror with the same apparent diameter as the full Moon but brighter, even if simply reflecting unintensified sunlight. If it reflected the driving laser or concentrated sunlight at Earth, it would be blinding, a dangerous weapon. Even at the distance of the nearest stars, a driving laser aimed at us would be easily visible at night as an intense one-color star, viewable simultaneously over a thousand-mile-wide swath of Earth. No such phenomena have been reported. Such a beam viewed from the side might also be visible on a dark night. The space between the planets, though a good vacuum by earthly standards, is actually lightly occupied by dust, which scatters sunlight and gives rise to the faintly visible Zodiacal light on clear and moonless nights. On such a night, a laser beam might be seen side-on as a band of one-color light. It would presumably have the same diameter as the sail it was aimed at.

DODGING LIGHT-SPEED BULLETS

All the above mentioned schemes that send a ship off at speeds over $1/100$ the speed of light have to contend with the fact just mentioned: space, though a vacuum, is not empty. In the vicinity of the Earth, human orbital debris numbers in the thousands. But natural debris, spread out around any star and even rarefied interstellar gas, are much bigger hazards. The typical dust and small asteroids orbiting the Sun have speeds in the range of 30,000 mph, extremely fast by our earthly standards. If they are in orbit with you, this is not too serious. But if your orbits are different and you collide with them then their speed relative to you can be almost that large (head-on, twice as large), and that is a serious issue, even though that speed range is only about $4/100{,}000$ the speed of light. Small rocks are a hazard for the occupants of the Space Shuttle and the International Space Station. Such impacts have been studied by several satellites, and based on all our experience are expected to be rare, at least for those craft. However, the photon ships and their sails are much larger and thus the probability of such collisions will be larger for them. When they accelerate up to $1/100$ or $1/10$ the speed of light, the penetrating and destructive power will rise by a factor of 10,000 to 100,000. Even single atoms will be deadly, penetrating radiation if encountered at near-light speeds. Thus any scheme for fast sub-light travel will require some kind of shielding, perhaps of two different sorts.

For dust and small rocks, a shield could be made by constructing water storage as the outer surface of the ship. But the photon sail would still be vulnerable. One might think of on-board detection systems linked to automatic laser cannon, to vaporize the rocks to atoms, which in turn might be managed by ionizing them with another laser so they can be deflected by a magnetic field "snow plow." This sort of shield would presumably also show up as a faint glow ahead of any such ship. Again, we have no UFO accounts to point to.

COLD SLEEP, HIBERNATION, AUTOMATIC CARETAKERS; THE ALTERNATIVE: GENERATIONS IN A COMPLETE AND SELF-SUFFICIENT ARK

All the sublight propulsion systems require trips of at least decades, and perhaps more likely centuries, to jump from star to star in our neighborhood, which is a typical spiral arm in our galaxy. The longest voyages on Earth have been in the range of several months. A trip of decades might be possible with everyone awake and active, but that is clearly pressing it in several respects. First, there is the amount of supplies needed to be carried on board, if wastes are not completely recycled. And the energy wasted as heat will need to be re-supplied somehow. Second, the physiological effects of not living in normal 1-g gravity will be serious if artificial gravity is not provided. Third, everyone will age significantly. Fourth, there are psychological issues of boredom and personality conflict in close quarters.

To each of these, there is an answer of sorts. The ship could recycle all wastes. To do this it will likely be necessary to provide a closed ecology. Such ecologies are now under test, but none have reached flight readiness. The energy for on-board light and heat and cooling machines will have to be supplied by the energy gathered en route and is a small fraction of the propulsion energy, though it will be needed thoughout the coasting phase if there is one. The ship could be spun to provide a sort of artificial gravity. Whether this will do the trick is as yet unknown. The aging process cannot yet be slowed down. The travelers could all be young adults at the start. The psychological issues can be treated as they arise, if properly trained personnel are on board and with proper medications. Careful ship interior design may help avoid some of these problems. If useful work existed for everyone throughout the voyage, that would help.

For any trip that takes centuries, all these effects would be much worse. There is a good argument to devise either a complete ark with a closed ecology in which the crew would live through generations (called in science fiction a generation ship), or some method of putting everyone or almost everyone to sleep in a low metabolic state for most of the trip. The former is somewhat closer to reality. Schemes for a whole enclosed ecology have been devised and studied on paper. Partial full-scale tests have been and are under way. Biosphere II in Arizona is a recent example of the latter, though it had problems.

The usual idea about metabolic suspension is a version of hibernation, or, as it is often called, "cold sleep." The primary problem with this is that the water in cells expands when it freezes, and therefore bursts the cell walls, which in turn collapse upon thawing. If a method of delivering antifreeze to every cell could be devised then perhaps cold sleep would be possible. But then all machines to support that would have to be perfect or self-repairing for the centuries of the trip. The technology for the latter can be visualized in principle. No laws of physics need to be violated. But such machines do not exist in practice.

No accounts of alien abductees mention ark-ecologies needed for generation ships. Nor are there reports of cold-sleeping aliens or their hibernation chambers. In a few cases, people report seeing humans or human-alien fetuses floating in aquaria. If the fluid is a metabolic substitute or preservative, it is just possible that these accounts point toward long sleeps and voyages. But no overheard alien conversations during abductions, themselves very rare, suggest such lengthy voyages.

If aliens had artificial gravity, matter replicators, artificially intelligent self-repairing machines, and a cold-sleep technique that worked through mind control, these problems could be solved. The ships could be smaller, and there would be no need for a complete ecology or hibernation chambers. Perhaps they have such things, but they remain undocumented.

THE EFFECTS OF EINSTEIN'S RELATIVITY ON THE VOYAGERS' AGING, THEIR MASS, AND THE VOYAGE

The various interstellar propulsion ideas outlined above have the possibility of achieving speeds up to about $\frac{2}{10}$ the speed of light. If they can make it past $\frac{3}{10}$, then the peculiar world of Einstein's relativity will reveal itself. As velocity approaches the speed of light, Newton's physics fails to explain what will happen. The major phenomena that would affect the interstellar voyage, not mentioned so far, are: the ship would become more and more massive compared to its rest mass, and on-board clocks of all types would run slow compared to those they left behind. The first effect would make each additional speed gain more energy-expensive. This would make on-board fuel schemes completely impossible and force one to employ gathered or beamed energy sources. For laser beam photon sails, adding each additional mile per second, would require more and more kilowatts. The increase would go up without limit as you approach the speed of light. This would clearly force an ultimate "practical" speed limit below 90 percent of the speed of light. The problem of collisions with space junk and interstellar atoms would be much worse and make shielding schemes difficult.

The second effect would help, however. All clocks, including the biological clocks of all living things on board, would slow and approach zero aging as the speed of light was neared. To people on Earth, it would seem as if the crew were hardly aging. For those on board, everything would seem normal, unless they compared themselves to those left behind, who would seem to be aging more slowly though in reality the return of the ship to Earth would show the opposite is true. The metabolic requirements of the voyagers would drop. Elapsed time on board would be only a fraction of "normal" time on Earth. In extreme cases, a crew could reach a nearby star in only weeks by ship time, but years would have passed on Earth. If they traveled to a star hundreds of light years away, for them the trip could take a year while for those on Earth their generation and several succeeding ones would have died (unless we discover life-extending technology).

In the closed world of the ship, all would seem normal, until one looked out the portholes. Then the trip would take on a strange aspect indeed. Those on the ship would see all of the starry sky in the front window shrink into a narrowing circle centered on the destination. The stars and galaxies would look bluer up to the time the blue shift converted all their radiation to ultraviolet, which would make them invisible to the naked eyes of the crew. Out the rear porthole the luminous sky would shrink toward the departure point, the stars and galaxies reddening until they, too, faded out. (Contrary to science fiction film renderings, the stars would not stream by.) A belt of completely black sky would fatten around the waist of the ship, a cosmic cummerbund that would grow as the ship approached light speed and would likely be quite disturbing to see.

So far, no alien abduction accounts report such phenomena. But hardly anyone has reported being taken to another star system.

4

LEAVING ALBERT BEHIND

TRAVELING AT WARP SPEED AND PHONING HOME FASTER THAN LIGHT

When George Adamski was welcomed aboard a flying saucer and then taken to its mothership, the aliens he met told him they were from the nearby star Vega. Some abductees report having been told by their abductors that they are visiting from a distant star, or even a distant galaxy. In any case, the aliens would have to have come a long way, and the length of their voyages would have to be years in the first case and tens of millions of years in the second, if they were limited to the speed of light and traveled directly from home to here, as discussed in Chapter 3. It would be much more fun to travel faster than light, or even to arrive instantaneously. In some abduction reports the aliens left the impression that they arrived from their home planet overnight or in a few months. Traveling light years in months or days means traveling at high multiples of the speed of light. Science fiction is full of references to "warp drives" and speeds of "warp 2" to "warp 10" (see the *Star Trek* films and TV shows). Is there any way that known physics can sanction such ideas?

SPACE WARPS AND WARP DRIVES

The idea of the space warp is, in a sense, pure Einstein. His General Theory of Relativity, published in 1915, makes clear that he considered matter and space to be inseparable. In contrast, Newton considered space to be a sort of empty stage, on which matter could cavort according to certain rules that he had discovered. Space did not affect matter, nor matter space. The force of gravity propagated instantaneously, so that the motion of one piece of matter would be felt immediately by all other matter in the universe. Einstein asked the question: Just how would this signal of gravity travel? And what is gravity anyway? Newton did not know the nature of gravity, as he explicitly stated in his famous book *Principia Mathematica*, wherein he derived his famous laws for its behavior:

> Gravity must be caused by an agent acting constantly according to certain laws; but whether this agent be material or immaterial I have left to the consideration of my readers.

Newton assumed that gravity travels with infinite speed. But when Einstein published his special theory of relativity in 1905, he assumed that nothing could travel faster than the speed of light and then deduced consequences, radical consequences, which could be and were confirmed experimentally. We discussed a number of these in Chapter 3. So Einstein could say with certainty that Newton's idea of gravity violated Special Relativity in its assumed infinite speed of travel.

When Einstein considered the idea that matter warps space (and time), he was able to picture a moving gravitational signal as a ripple in space and time. He then calculated the speed of the ripple based on the bending of space and the properties of space, and he found that the speed of gravity signals was exactly the speed of light in a vacuum. Under his theory, there was no contradiction.

To picture the influence of matter on space, consider the following analogy: visualize a sheet of rubber stretched taut inside a frame so that it is flat. Place a very small and light ball bearing on this membrane and give it a push. The ball will roll in a nearly straight line until it encounters the frame.

Next, place a bowling ball on the sheet. It will sink into the sheet and create a funnel-shaped slope in the sheet centered on the ball's center. Leaving the bowling ball there, roll the ball bearing across the sheet and it will not travel in a straight line. It will travel a curve. If you give it enough speed and aim to the side of the bowling ball the bearing will follow a curve around the funnel and out again. If you aim the bearing just right it will go into an orbit around the funnel, generally an elliptical orbit, just like the planets around the Sun. If there were no friction, the ball bearing would continue in orbit for a long time.

The bowling ball warps the sheet, which represents space, and the bearing stays in orbit because of its momentum and because it follows the local curvature of "space" created by the bowling ball. Einstein proposed that all matter bends space around itself. The larger the mass of an object the larger the surrounding dip or funnel in space it creates, and the closer to the mass you are the greater the funnel's slope (curvature). Thus, pieces of matter follow a curved path when in the presence of other pieces of matter. We can call such bending of space a space warp.

The problem with the analogy of the bowling ball on the rubber sheet is that it is too easy to visualize. It is a case of a flat surface distorting into the third dimension, which is a piece of cake to imagine. But in real life the situation is much harder to imagine. Matter distorts space all around itself, fully three dimensionally. We can think of a blob of matter, like the Sun, centered at the bottom of dozens of overlapping fabric funnels, all of the same shape, which open outward and face in all directions. No matter what path an asteroid or comet takes, it will skirt the edge of one or another funnel and follow a curved path across it, rounding the Sun.

When any piece of matter moves in space, the local bending it creates moves with it. Its motion causes its local bending to shift and gives rise to ripples in space-time that spread out like ripples from the wake of a boat spreading across a lake. These ripples are the signal of gravity, and they are what move at the speed of light. In modern terms, we can call these gravitational waves, or, in their particle form, gravitons.

Any mass will respond to the curvature of space—that is, the presence of another mass in its local vicinity. If the predominant curvature is not centered on itself, it will move in the direction of greater curvature. The mass will seem to "fall" into the direction of greater curvature, accelerating as long as the curvature increases. Near a body like Earth, the curvature centers on the Earth's center and objects near its surface fall toward the center, accelerating as they go. This was pointed out by Galileo back around 1600. Falling objects are, of course, stopped by the Earth's surface, good old *terra firma*.

But our business is alien ships' ability to travel between the stars and galaxies at speeds greater than that of light. If we could warp space to really great curvatures in a controlled way in the direction we want, we could accelerate a ship in that direction very efficiently. It would be like sending it falling onto a planet or star. We would want the ability to bend space in front of and behind the ship, in complementary fashion. If we could do that, we could, in effect, shorten the distance to be traveled in front of the ship, cramming the space together in front and letting it spread out behind the ship. The late science fiction grandmaster, Frank Herbert, used this conceit to move his ships quickly across vast interstellar distances in the novel *Dune*. He called it "folding space." This would be somewhat like dragging yourself across a bed by grabbing and bunching fabric in front of you and releasing it behind as you pull yourself by. In any event, doing this would effectively shorten the distance to be traveled.

How could this trick be pulled off? Perhaps by dangling a very massive but compact object on a pole in front of the ship, using its space warp to carry your ship along. The problem is you have to also propel that object, which requires energy. It would be better to find a way to create a steep warp without mass, in effect an artificial black hole. (Another way of looking at the problem is that we would be creating the grin of the Cheshire cat while omitting the cat.)

A black hole is itself a curious beast. As you likely know, it is created by the formation of an object so dense that at its surface the velocity of escape is greater than the speed of light. Because nothing in the universe travels faster than light (see Chapter 3), nothing, not even light, can escape. Contrary to what you might think, any amount of mass can form a black hole, as long as it collapses to achieve a small enough radius and a large enough density.

Scientists believe that in nature black holes have formed in at least two ways. The medium-sized ones form when a star ten or more times heavier than the Sun finishes its life in a supernova explosion. The inner core of the star, at least three solar masses, collapses in the explosion. While the stellar core starts with a radius of something like 100,000 miles and when it implodes to a radius of just ten miles it becomes a black hole. That represents a drop in radius of a factor of 10,000, and a rise in density by a factor of about a trillion. These kinds of objects must be scattered around our galaxy as the result of the rapid evolution and death of massive stars. A few cases are now fairly well observed.

A second and altogether larger type of black hole may be formed at the dense center of a galaxy when stars and gas are so tightly packed that they collide in large numbers, though they may not go supernova. In many galaxies, including our own, there are huge but non-luminous central masses made apparent by their extraordinary pull on stars orbiting near the center. In some cases, there are huge fountains of hot gas being forced out of the center by energies way above normal. These cases all point to the presence of black holes, with masses in the range of 100 million solar masses. In other words, millions of stars have been eaten by these supermassive black holes, perhaps over millions of years.

Every mass that becomes a black hole does so when it compactifies to a certain radius, as mentioned above, at which the escape velocity is the speed of light. The mass keeps on collapsing, but the surface with this radius remains as it was and is called the event horizon. The larger the mass the larger the horizon. For individual stars, the radius of the event horizon is about ten miles. For a 100 million solar mass black hole, the event horizon will have a radius larger by the same factor as the mass ratio, in this case a factor of 30 million or a radius of 300 million miles. That is about three times the distance of the Earth from the Sun, or about 24 light minutes.

There may also be micro black holes, formed when the universe was very dense and hot, just after the Big Bang. If these have masses about a 100-millionth the mass of the Sun then they have a radius of a 100-millionth of a mile, or about six ten-thousandths of an inch. At this scale Stephen Hawking predicts from quantum theory that the hole will be unstable and actually evaporate. If you had one captive, it might be a great though lethal source of energy, as it would evaporate by radiating deadly gamma rays.

In any event near any black hole of medium or smaller size, the curvature of space becomes extreme, just the thing for a ship to "fall" toward. If we could create that curvature without the mass, control it at will so that it was always ahead of the ship, but never close enough to suck the ship in permanently, it would be a warp drive. Space would be compressed ahead of the ship and expanded behind.

What might we see if we were in the ship looking out, or standing in its wake as it departs for its voyage? Because the ship travels in a sort of warp pocket, it moves locally (inside the pocket) at less than the speed of light. Any light that entered the pocket at departure is trapped with the ship and the view out front and back looks normal until all the light is absorbed and the sky turns black. When the ship drops out of warp, the sky reappears and, if its speed is well below that of light, the sky looks normal. If the ship has traveled to another star system, the stars will be seen in different positions than we see from our solar system when the sky reappears. The constellations we know will be exchanged for new ones.

A watcher in the ship's wake would likely see the sky around the ship waver like a desert mirage as the ship accelerated away. If it happened at night or in space the stars in the sky might appear to draw in around the ship, and as the stars crowded together this area might for a time look like a bright ring of sky. As the ship became more distant, this luminous doughnut would shrink in radius. The ship would get redder and fainter the faster it traveled until it and the shrinking luminous belt faded out.

Does any of this sound like anything reported in UFO sightings? Not exactly, though in some cases departing ships are reported to disturb the stars, leaving a sort of wake. But in most cases, these are departures from inside our atmosphere and in that case simple heating from a hot drive would create currents in the air of an ordinary sort that would create the optical wake. It would actually be unexpected for an alien to turn on a warp drive inside the atmosphere of a planet, or for that matter anywhere within thousands of miles of a planet. As pointed out in *Star Trek* and in many other sci-fi movies, a powerful warp drive, with its extreme curvature of space, would do major local damage if turned on too close. A ship would be more likely to use something else to get off-planet.

Another feature of the warp control is that inside the ship there would be no sensation of acceleration, as the mass of the ship and its contents would be at rest in its local pocket of space. With slight modifications this could also serve as a tractor beam. In addition, because the ship is locally traveling below the speed of light, clocks on board will run at normal speed, and the ship's occupants would feel none of the time dilation predicted by Einstein's special theory of relativity, as discussed in Chapter 3.

There is yet another advantage to warp control. The warpage could be so great that light would travel on steeply curved paths near the ship. Those paths would be so curved that light and matter would travel around the ship. This is tantamount to a deflector shield.

HOW WARPED ARE WE, ANYHOW?

So the aliens have the warp drive and they can turn it on anytime after leaving Earth's vicinity. What would warp 10, *Star Trek*'s top speed, do for them? According to the show's definitions, warp 9.6 is the highest normal speed and it is 1,909 times the speed of light (in *Star Trek* warp 10 is an unattainable infinite speed). At 1,900 times the speed of light a trip to Vega, about ten light years away, would take 10/1909 of a year or about 1.8 days. This makes Adamski's visitors' statements more reasonable. A trip to the far side of the galaxy, about 100,000 light years, would take 50 years. One really could travel the galaxy, though the trip would be on the long side. What about intergalactic travel at warp 9? The Milky Way's closest neighbor is the Magellanic Cloud, at a distance of 160,000 light years. The trip would take a human lifetime even at warp 9. The next nearest destination, M31 in Andromeda, is 2.5 million light years away and getting there would take 1,300 years.

There is one nagging question to answer: What, if anything, could produce a strong warp and how much energy would that require? The answer appears to be, based on quantum mechanics, that the warp requires a kind of energy called negative energy. Just what this is we will talk about further when we consider wormhole technology. But how much energy will it take to make and manipulate this energy and use it to fashion a strong warp? For estimating purposes assume that aliens could create a black hole the size of a spaceship 1,000 feet long (really big). Such a hole would warp space enough in its near vicinity to act as a warp shield and distort space as much as a warp drive, though it would have the wrong shape. Because a star 10 times the mass of the Sun turns into a black hole when it shrinks to 10 miles in diameter, the 1,000-foot black hole would be created by a mass equivalent to about one-fifth the mass of the Sun. The energy equivalent of that (according to Einstein's $e=mc^2$) is equal to the energy radiated by our Sun over its whole lifetime. Needless to say, there is no *known* source that could provide this energy requirement in any manner, much less on board a ship.

Space warps do not seem ruled out by the laws of physics. If it turns out to be possible to construct them, they could serve not only as warp drives but as warp shields and tractor beams. The problem is finding the incredible amounts of energy to run them. For this, physics today has no answer.

THE ALICE-IN-WONDERLAND OF FALLING INTO BLACK HOLES

In the process of describing warp drives, we used the idea of black holes in a general way without really describing their innards or what would happen if you fell into one. It is often said that black holes might be the way to faster-than-light travel, but that is not really true. To achieve that goal, the general theory of relativity points either to warp drives or to wormholes. We just discussed the former. The latter are related to and sometimes confused with black holes, but are not the same thing. We will discuss them next.

Black holes, as described above, form during the destruction of a massive star. The star's core falls through the event horizon and continues to contract. What happens to it after that? Often it is asserted that the matter will continue to contract until it reaches infinite density, in what is called a singularity at the center of the hole. If that is true then general relativity and all other ordinary theories of physics will break down. They can not handle infinity. However, quantum mechanics suggested to Stephen Hawking that the singularity might not occur. Instead, the center of the hole might hit a minimum size and stop contracting. The new string theory (see Chapter 6) predicts just that. Relativity is saved even deep in the hole. The matter stops its collapse.

Meanwhile, what is happening outside the event horizon? Far enough from the event horizon, say beyond 100 horizon radii away, the curvature of space is so slight that it seems as if it's normal gravity from matter of the same amount as fell into the black hole. Looking toward the hole, if there is no infalling matter, the hole looks black.

Turn on the drive and steer nose down toward the event horizon of the stellar black hole. You see the blackness approach. Looking to the rear, toward outer space, the stars and galaxies appear to get bluer and bluer until they become ultraviolet and fade out of our eyes' range of sensitivity. As the ship approaches the hole and the blue shift occurs, the sky becomes distorted behind you, and the stars and galaxies appear to crowd together, contracting into a smaller and smaller patch of sky over your ship's tail, surrounded by a growing ring of blackness. You will enter the hole faster and faster as you fall in and flash through the event horizon. After passing through, the sky should look like a point of extreme ultraviolet or X-ray radiation, if you are around to see it. But will you be?

At the same time that the curvature of space near the stellar black hole starts to severely blacken the sky above your ship, it also begins to be large enough that gravity pulls harder on your head, nearer the hole, than on your feet, which are farther away. This is an extreme case of the tidal effect. As you get close to the event horizon you and your ship will be torn apart.

An observer in a ship sitting outside the vicinity of the black hole will see your ship do something rather unexpected and different from what you see. As it approaches the hole, the ship will appear to redden more and more until the light it radiates shifts into the infrared and fades from sight, out of the sensitivity range of human eyes. At the same time, the ship will appear to glide to a stop right at the horizon. The message of its passage to the inner world of the hole will never quite reach the outside world. This is simply the opposite effect of relativity to the blue shift seen from the ship, looking backward to the outside universe.

A SURVIVABLE VISIT TO A BLACK HOLE?

All the above statements apply to visits to black holes that have arisen from stars that died alone. But the theory of black holes shows that if the mass in the hole is bigger, the hole's event horizon is bigger, and paradoxically the tidal effect at the event horizon is smaller. If the mass in the hole is large enough, a ship and passengers could pass through the horizon without being ripped apart. A black hole with 100 million solar masses inside would offer such a possibility, and as mentioned above, astronomers believe that such an object exists at the center of our galaxy and also at the centers of many other spiral and elliptical galaxies. Such a black hole would have a radius about three times the radius of Earth's orbit (about half the size of Jupiter's orbit).

In this case, your spaceship would pass through the event horizon without major mechanical incident. The universe outside would seem to you to fade to blue and contract to a point overhead, as described above. The event horizon would still signify that the escape velocity from the hole is at or above the speed of light, so no light or radio signal you sent outward through space could escape.

But wait: If there is no mechanical destruction at the horizon, could you trail a coaxial cable behind your ship to an auxiliary communications pod outside the event horizon and use that to send home a report of conditions inside? Alas, no. The speed of light still rules in the cable, and it is below escape speed. But wait again: The blue glow in the water surrounding nuclear reactors signifies that matter, in this case electrons, can travel faster in water than light travels in water. This is called the Cerenkov effect. (The top speed of the electrons in water or other matter still must be less than that of light in a vacuum.) Can this be used to advantage, using modulation of a beam of high speed electrons to penetrate the horizon inside a water-filled tube, in place of the solid cable? Alas, no again. Your voyage can be made, you will learn the secrets of the interior of a black hole, but you will not be able to tell the rest of us waiting outside what you find.

All this discussion of black holes has led to a very deep dead end, as far as faster-than-light travel and communication. We expect no aliens to arrive with true tales of secrets of singularities or the "bottoms" of black holes. But it has prepared us to talk about another possibility for their arrival at speeds that make cross-galaxy travel feasible. This possibility is the creation and use of wormholes.

Tunnels in the Sky

Think again of our black hole analogy to a bowling ball sitting in its own pit on the surface of a stretched fabric. Now consider what would happen if the fabric were in a c-shaped frame so that it bent under itself. Suppose a second black hole could be created on the underside of the folded-under fabric, in a location directly below the first hole. Further, suppose the second hole points up toward the first and that they meet in the space inside the fold, between the two parts of the fabric, and then somehow the two dimples join up to make a continuous, smooth, open tube. You now have an analogy for a wormhole connecting two locations in our universe. But can this be done? And what would this get us, or any spacefaring aliens?

Well, it would get us a travel short cut of great effect and also might very well get us time travel of a sort. The short cut occurs if folds can be made in the space of our universe. If one end of the wormhole is local, near the outer edge of our solar system, and the other end is near the bright stars Vega or Betelgeuse (25 and 500 light years away, respectively), then if we can jump through the wormhole to travel to these stars, the distance to them through the wormhole in our space is zero even though light will take 25 or 500 years to travel to us from them in our universe. (Please note that the hole itself, the tube, is not in our universe and is likely not in any other universe. You conceive of it as being in "subspace" if you like.) Thus, by the ship's clock, the trip through the hole would be instantaneous.

The travel time is then drastically reduced to just the time spent to go to and from the wormhole in normal space, at speeds below that of light, on either end of the trip. This time would still not be negligible if we or aliens are limited to the best speed allowed by ordinary rockets. To travel to Jupiter's moons, for example, takes years, even using a gravity assist (the "slingshot effect") by being flung around the Moon or Mars. But it would not take decades or centuries. Quite an improvement!

If a wormhole could be created and we could travel to it and look inside, we would expect to see the whole universe as we would see it from the other side, our back to the hole. Looking from the other side towards home we would see the whole universe from the solar point of view, crammed inside the perimeter of the wormhole.

We should also note that, though such things are not part of any of the standard UFO and alien abduction narratives, the popular idea of the stargate is functionally equivalent to a wormhole. Perhaps the first clear depiction of a stargate is to be found in Robert Heinlein's classic novel *Tunnel in the Sky*,

A tunnel in the sky? A wormhole appearing? This strange "object" was shot by a field technician at a Colorado mountaintop weather station. Photo: copyright 2000 Corbis-Bettmann/UPI

and it is seen today in the recent movie *Stargate* and the TV series *Stargate SG-1*. In the movie and TV show the gate is shown operating from the surface of Earth, linking to the surface of other planets. This might not be wise for two reasons: First, the mouth of the hole will be a location of great space curvature. That may be damaging to its immediate surroundings. Second, as Heinlein pointed out, planets are constantly moving and, celestially speaking, tiny targets. The aiming problem would be almost impossibly severe.

Severe, that is, unless a gate were transported by ship to the destination, from where it might signal back, allowing easy detection, alignment, and connection. But wait! That defeats the advantage of the stargate in two ways: It takes years to deliver the gate the long way through space, and the signal back takes years to return! However, we can get around this if we have a spaceship-sized wormhole operation. Then we open the wormhole near the destination planet, deliver the gate to the planet's surface, and then send the gate-alignment signal back through the wormhole. After alignment, we can dispense with spaceships and simply walk through the gate to our destination—assuming always that the gate does not destroy its surroundings.

By the way, stargates operating like this will almost certainly not look like the one on *Stargate SG-1*, that is with a roiling opaque cloud or flexible membrane in the opening. Instead, you would see an extreme fish-eye view of the planet's surface on the other side. The TV image is, however, beautiful and suggestive.

VISITING THE PAST

What would happen to our sense of time as we jumped through a wormhole? Here is where time travel comes into play. It seems likely that there are two time travel effects to consider. First, as pointed out by relativistic astrophysicist Kip Thorne, if the wormhole is somehow anchored to space at our end, perhaps by a local space warp, but free to roam at the other, a time slip can arise. The reason is that the far end of the hole will be traveling with respect to its local space. If that speed is greater than about 50 percent of the speed of light then Einsteinian time dilation will be important at that end of the hole. But "local space" at the hole's far side is connected to and thus is part of the fabric of our universe, and, however distant, is thus connected to our local reference frame. This means that the time dilation effect is not only with respect to the far side of the hole but also with respect to the origin (our) side. If we put a clock through the hole, we should see it run slower than its twin left on our side of the hole. If the wormhole length does not change during this exercise and we are inside the hole looking out, clocks at either end will appear to run at the same rate. If the far end of the hole is then returned to its starting point in space, and if for us the process takes a week, but for the people on the far end it takes a day, we will see their clocks and calendar register only an advance of one day while ours register the whole week. So by our clock the hole-wandering might have started on the 14th of June and it may now be the 21st of June, the solstice on Earth, but at the far side it would only be the 15th. If we then travel again through the hole we would be traveling back in time!

There is a second aspect to the time travel possibility. Whether or not both ends of the wormhole are anchored to their local space, a jump through the hole would mean we would arrive at our destination before light from Earth would arrive the "long way," through normal space, to portray our moment of departure. We would arrive before we can be observed to have left. If we return home immediately we will also be back before the light signal of our original departure arrives at our destination. If we put a telescope and TV camera through the hole looking toward Earth the long way and feed the picture back through the hole, we can jump through the hole, do our business on the far end, jump back, and after the passage of years finally see the image of the ship leaving Earth from the comfort of our living room—a *non*-instant replay.

We can deduce two interesting consequences from these ideas. First, the time travel possibility is somewhat limited. Travel into the past by wormhole can only reach times *after* the wormhole is established. In other words, we or the voyaging aliens have to be alive to travel into our past, and the trip can't take us to a time before we left. This would seem to guarantee that no one can go back and kill his grandparent before his parent is born, and thus there is no potential to cancel yourself out. That neatly avoids a major and well known paradox of time travel. Second, if wormhole travel, or for that matter space warp travel, is widespread in the galaxy, then there should be a multitude of ghost ships—images of the departing ships—floating around the universe, trailing their originals by many years and traveling at light speed. Perhaps there is a unique signature to these we could look for. Perhaps they are part of the general background brightness of the night sky.

ENGINEERING WORMHOLES

We have so far omitted any discussion of how to construct a wormhole, or any estimate of what it would cost in energy terms to operate one. Though science fiction is rife with phrases like "Jump through the wormhole" and "Open the stargate," the creation of a wormhole will likely not be easy, nor will keeping it stable once it is open.

First, how do we or any alien open the hole? Quantum theory suggests that nature may create wormholes of subatomic dimensions constantly, as virtual particle pairs flash into momentary existence in our universe, coming from nothing, and then annihilate away. The uncertainty principle of quantum mechanics rules the mass, speed, and time of existence for these virtual particles. The particles and their holes disappear immediately in almost all cases. This mechanism keeps the vacuum of space seething with energy and random creation/destruction events.

If one of these holes could be isolated and grabbed, and then enlarged, that might be a start. This will require energy, and some actual structure made of matter might have to be inserted to stabilize the hole against the stress of space that tries to collapse it. Picture finding a pinhole in one wall of a balloon and inserting a hollow needle there that tapers to a point. Then, picture pushing the needle in, dragging the balloon skin in, enlarging the hole, continuing to the opposite wall of the balloon, puncturing it and seamlessly joining the skin at the second puncture to the skin dragged across from the first.

What material could serve to function as the hollow needle, supporting the wormhole against collapse and with space inside to allow spaceship transit, or in the case of a planetary stargate, people to walk through it? Kip Thorne has again proposed an answer. The calculations of general relativity he has carried out suggest that the stress could be met if the matter inserted is what he has named *exotic* matter. A major property of this matter is that it would have negative energy (compared to the positive energy of ordinary matter). It might also generate a repulsive rather than an attractive gravitational force. Another way to say this is that exotic matter will create a negative space curvature in its immediate surroundings. Recall that matter warps its surrounding space, and that at the event horizon of a black hole space is extremely positively curved because ordinary matter at extraordinary density is inside. The mouth of the wormhole will also be positively curved. Thus it seems logical to insert a wall of negative curvature to hold it open, and Thorne's exotic matter with its negative space warping will fill the bill (and the hole).

There might be another way to design the entrance to a wormhole. That is by creating a black hole with an event horizon in the shape of a spinning torus (doughnut), instead of a sphere. Inside the torus hole, space would have negative curvature, and it might be possible to create the wormhole mouth there and have it be reasonably stable because of this geometry and the centripetal force of the spin. To create the black torus would require a strange and perhaps impossible-to-achieve distribution of matter. Or perhaps it could be created by squeezing a spinning ring distribution of matter to super densities. This can be reliably guessed to be a tough project, because the international controlled fusion programs have attempted almost the same task in the form of a gas plasma magnetic pinch for the last 40 years on a much less extreme scale and success is still elusive.

You might ask whether any stable hole with dimensions of a spaceship or smaller would allow passage safe from extreme tidal effects, since space will be highly curved at least at the inlet and outlet. The precise size and form of the tidal effects will depend on the wormhole's cross-sectional shape and on the path taken through the hole. But for cross sections with circular or regular polygonal symmetry, whether the hole is lined with ordinary or exotic matter, a path down the wormhole axis should equalize and thus remove all tidal effects.

Our engineers are now left with the same question that was tackled at the end of our discussion of warp drives: supposing wormholes can be created, how much energy will that take? There are currently no really good estimates for the energy required. In their place, we have to fall back on the estimate made for warp drives, and we get the same answer. That is, assuming that a black hole the size of a large spaceship, 1,000 feet across, is the basis for the calculation, the energy required will be about equal to the total ouput of the Sun over its whole lifetime. This is, again, a frustrating result.

ON KNOWING YOUR DESTINATION AHEAD OF TIME

There is another problem with wormholes that needs some thought. The wormhole engineers will probably be unemployed if they cannot open a wormhole to specific destinations. This might not seem like much of a problem, but it is. There are at least two important impediments to aiming at a target: lack of a hyperspace road map and long distance precision. Of these, the first is more difficult and serious.

The alien wormhole engineers would ideally want to open a wormhole between Vega (for example) and our solar system. To do that as we have described, they would have to create a black hole or open wide a natural wormhole. They would then have to line it quickly with exotic matter. The holes will have their extension or length outside of our universe in a higher dimension, or hyperspace, and with a particular direction in hyperspace. However, until theory or observation advance, there is no way to create a map that provides correspondence between a direction and distance in hyperspace and any destination in our universe.

Bob Shaw, in his novel *Night Walk*, suggested a way for making a map that depended on two unlikely assumptions. The first was that wormhole entry portals had been discovered more or less accidentally and put to use by sending robot probes through them and then waiting for their reports to come back by radio the long way. Because the nearest destinations were at nearby stars (which are light years away), the waiting periods were a minimum of several years. Robots to much more distant portals could report, but their messages were not going to be received for centuries or millennia. The second notion was that one could detect the paths of spaceships making wormhole transits through hyperspace if one were sitting in hyperspace to observe them. I don't want to spoil the novel, which is excellent, by revealing more detail. Suffice to say that with these two assumptions a map could be made.

There are more brute-force alternatives. If it turns out to be easier to create very short wormholes, ones that connect to locations only yards apart in our space but that can be stretched in length once opened, then one could keep one end in the home stellar system and put the other end on board a space ship and carry it to the desired destination. If the ship is limited to sublight speeds, it will take decades to build up a wormhole network within the space near to the home system, since stars are, on the average, about 10 light years apart in our arm of the galaxy. Of course, decades or centuries are blips on the cosmic timescale.

If the ship has warp drive then the process would be nearly instantaneous, except for sublight travel in close-in space. However, if one has the warp drive, why use wormholes? The only reason I can see is for the convenience of walk-through, planet-to-planet travel, as in *Stargate*. In that case, the engineers would transport the gate by warp drive and then land on the planet and install the gate.

One could possibly try opening wormholes in random directions and hope that by looking through the hole one could identify the location of the far end by the appearance of the sky visible through the hole, and use that information to adjust your aim. However, there is an ethical problem with shooting blind, because there is no guarantee that the far end would open in empty space, far from a place that could be damaged by the event. And of course, there is the risk that the far end could open into a black hole or at the outer surface of a star as it goes supernova, in either case posing a severe risk to the engineers and their home system.

ZERO POINT ENERGY AS FUEL

Based on the energy considerations outlined above, it might seem hopeless to expect mastery of either space warp drives or wormholes; this makes it unlikely that any aliens have control of these technologies either. There is one development on the horizon that might, possibly, lift this pall of pessimism. It is an energy source that pervades space that was briefly mentioned above: the quantum fluctuations of the vacuum momentarily create and destroy subatomic particles and photons. Tapping this evanescent seething sea of energy could turn out to yield energy anyplace in the universe. Sometimes this energy is called the *zero point* vacuum energy.

The suggestion that one could observe such energy hiding in the vacuum of space, other than by the conventional observation of particle pair creation events, was made by physicist Henri Casimir back in the 1940s. He suggested that the vacuum could be thought of as being filled with virtual photons of all different wavelengths (and thus energies). He predicted that if you create a vacuum between two facing reflecting plates, then there will be a tiny force created that will push them toward each other. The reason is that the reflecting plates act like a resonant cavity. The virtual waves with wavelengths that just fit the cavity opening, that is for which the cavity is a multiple of a half wavelength (½, 1, 3⁄2, 2 etc.), are reinforced and resonate inside, as in an organ pipe. All other wavelengths are destroyed. But outside the cavity all wavelengths exist, so the radiation pressure from them pushing in is greater than that of the inside waves pushing out. It was not feasible to do the experiment in Casimir's time, but recently with new technology an experiment showed the effect. Thus there is direct evidence for the existence of zero point energy everywhere.

What is not known is just how much zero point energy is locally available. Available in this case means how much in theory can be extracted and used. There might be a lot of energy packed into each point in space, but it might be at too low an energy level to be used. There is a close analogy here to gravitational potential energy. If you lean out a second story window, you have a certain gravitational potential energy that will be liberated if you fall to the ground. It is enough to knock you silly or seriously injure you in some cases. But you have a much larger potential energy with respect to the center of the Earth. Yet it is not available, and nothing will come of it unless arrangements are made to drill a hole directly under you down to the center of the Earth. Then, look out! At present, the potential energy level of the zero point energy (part from gravity and part from photons) is not known, and no experiment has been executed to determine it. We also know of no way to extract it on a large scale.

The second consideration here is a serious environmental impact-type question: Would large-scale extraction of zero point energy affect the vacuum in a permanent and/or undesirable way? Here again, we really have no clue. If the fabric of space-time depends on the zero point energy, extracting the energy on stellar scales (remember we estimated we might need the Sun's lifetime output to construct a usable warp drive) might have severe and wide-reaching consequences.

SIGNALING THROUGH WORMHOLES: WHAT IS THE RECEIVER?

Suppose wormholes are too energy expensive to construct on a spaceship scale. Perhaps we could construct miniholes instead. Or maybe the larger ones can be made to work. We can ask in either case whether it is possible to signal through the wormholes. It probably is simpler to do that than to send ships through. But is it dead simple? Not quite. We have discussed that light entering a black hole is blue-shifted going in. If it came out from the same hole (though we know it can't), it would be red-shifted back to its original wavelength. Going into a wormhole, from which light can emerge, it will probably be blue-shifted somewhat, and assuming the far side of the hole has the same dimensions and space curvature, be red-shifted back to its original wavelength as it emerges on the far side. But there is a complication. Kip Thorne seems to have shown that the exotic matter needed to line the hole exerts its negative gravity in such a way as to greatly defocus the light as it passes through. Aiming a tight beam right down the axis might, for symmetry reasons, avoid this effect. If true, a bright beam could be sent through, but then the receiver has to be right on the axis of the hole to receive it. This imposes the need to be in the right place, or the engineers have to steer the opening of the hole to point it in the right direction. This is probably feasible in either case. A tiny piece of good news!

If we have to, we can send a receiver by sublight ship the long way and then open a wormhole and signal through it. This would take a long time to set up, but afterward colony ships could be in instant touch with home. News would travel fast, even if people could not. The colony would be on its own physically, but advice and psychological support would be immediately forthcoming.

A spacefaring alien civilization might spread out slowly among the stars, taking centuries to create an inhabited bubble of space 100 light years across, setting up outposts in each new system, perhaps including ours, and always remaining in close cultural contact with their home planet as they go. Perhaps such an arrival has taken place in our solar system. But none of the UFO or other encounters of any kind have yet hinted that that is the case.

TACHYONS—QUICKSILVER MESSENGERS

The Special Theory of Relativity consists, in part, of equations that have both positive and negative signs. About 25 years ago, it was proposed that the negative sign might signify a kind of matter that saw the speed of light as a speed limit, just like our ordinary matter does, but as a *lower* limit, not an upper one. Particles with this property would have their lowest energy state at infinite speed and their highest at the speed of light, which they could never quite reach. These particles were named *tachyons*, which is Greek for speedy particle. This idea was taken seriously for a while, and experiments were designed and built to detect such particles. A few careful searches failed to find any, and the enthusiasm for the concept has lapsed. If they had been found, they could have been used to violate the usual chain of cause and effect. That is one reason the original idea was not received with universal enthusiasm among physicists.

CAN EFFECT PRECEDE CAUSE?

In much of this chapter we have worried about faster-than-light travel and its relation to time travel. I alleged that the properties of wormholes allowed time travel but avoided time travel paradoxes, which might arise, for example, if you can murder your grandparents before your parents are born. Such paradoxes arise if you can travel into your own past. Wormholes apparently don't allow such things to happen. Tachyons would. These problems are part of a general range of possibilities that come under the heading of a cause preceding its effect.

There is a great horror of such causality violations in the physics community. The feelings are so strong that a number of scientists have proposed rules or laws that universally forbid causality violations. Thus, when a proposal is made for some new idea or phenomenon that appears to jump past causality, these scientists reply curtly that it is ruled out precisely for that violation, independent of any other considerations. No new physics or deep analysis is involved in such pronouncements; it is simply that the rule must hold for the universe as we know it to make sense.

5

"Beam Me Up, Scotty."

MATTER TRANSFER BEAMS FOR LOCAL AND LONG-DISTANCE TRAVEL; THE "FREEZE AND TRANSPORT" BEAM IN ALIEN ABDUCTIONS; THE ACCOUNTS OF LINDA CORTILE AND OTHER ABDUCTEES

In many abduction accounts the abductee reports being transported to an alien ship by a beam of light, as if weightless, in total silence and paralyzed. In many cases, this process seems to waft the abductee right through solid surfaces as if they were not there. No accounts mention any explanation of this process by the abductors, or the characteristics of any machine that might make it happen. Usually no outside witnesses to this transportation mode exist, but in two cases—the Linda Cortile and Travis Walton cases—there were multiple witnesses. In the Cortile case, these witnesses did not know the abductee and presumably had no relation to her. In the Walton case, the witnesses were his friends.

In almost all such cases, the abductee is aware of the outside world, and the witnesses confirm that the abductees do not dematerialize. They are visible and look normal, if immobile, throughout. This rigid mobility is more like the behavior we would expect from the use of a tractor beam (as discussed in Chapter 4) than anything else. Presumably a tractor beam would exert its force on all atoms of something simultaneously, which would tend to retain the object's shape, causing the object to look rigid. The problem with this picture is that the envisioned behavior of tractor beams does not include the ability to pass through, or carry something else through, solid walls.

Contrast this picture with that of the transporter in action on *Star Trek*. In that scenario, the traveler appears to dissolve smoothly at departure and re-corporealize smoothly at the destination. If this happened in an alien abduction, the abductee would presumably see the original surroundings fade out and the new ones fade in, with no sense of motion in between. They would be able to move freely as this took place, as long as they did not let a body part stray out of the zone of transport. To the witnesses, this would look just like a scene from a *Star Trek* episode.

The visuality of this is probably what one would expect from the process explained below, called quantum teleportation, with one difference. In that process, there would have to be a sacrificial person at the destination who is transformed into the new arrival. This would complicate things a bit, would it not?

MASS TRANSFER IN BULK

The transfer of matter in bulk would constitute another scenario. In such a case, a whole object would blink out in one place and "simultaneously" appear all at once in another. At least that is a popular sci-fi scenario. It is, of course, possible that the object might merely travel at the speed of light to its destination rather than at infinite speed.

There is no really good possibility for such an effect. A tractor beam coupled to an artificial wormhole or to an artificial space warp might seem to operate like this, if it could be energized very rapidly. However, for any such "instant" transport, there would likely be observable side effects not mentioned above. A primary one would be the thunderclap that would accompany the departure, as the air surrounding the object collapses into the region of perfect vacuum created by the object's removal. At the arrival, a small explosion would also be heard. And there would be problems for an object trying to materialize into a space already occupied by air atoms. Perhaps to avoid this problem, the destination would need to be inside a vacuum chamber. Of

This photo was taken in Taormina, Sicily, in 1954. These men are watching the strange objects in the sky, which appear to be saucers flying upside-down. Photo: copyright 2000 UPI/Corbis-Bettmann

course, if that is the case, the chamber would have to be transported ahead of time by some other means. If that means is at sub-light speed, the establishment of the chamber network around the galaxy would take many millennia. The galaxy, of course, has been around for eons, so aliens have had plenty of time to stake out their territory.

DISASSEMBLY DOWN TO ATOMS AND REASSEMBLY AT THE DESTINATION

Any non-bulk transport system avoids the sonic boom problem but has some far more serious issues. The primary one arises out of the very fact of atom-by-atom transport and the gigantic number of atoms that must be handled in the process. For inanimate objects, it is difficult enough. Any solid the size of a human being contains about 10,000 hundred million hundred million hundred million atoms (about 10 to the 28th power). Their relative positions must be detected and preserved and then duplicated in reassembly at the other end. The requirement may be relaxed a bit for a material that is either purely one element or molecule or for an amorphous mixture. In those cases, the reassembly requires only that all or almost all of the atoms are reincorporated, but because all atoms or molecules of any one type are interchangeable, and because there is no order in amorphous materials, a huge number of arrangements of individual atoms will be equivalent to one another. The most difficult inanimate reassembly problems would arise for crystals with dopant (impurity) atoms in a particular crystal arrangement, a solid with a complex compositional gradient, or a computer chip with data recorded in memory.

Consider, now, the time for disassembly and reassembly. If one atom or molecule is removed every femtosecond (10 to the minus 15 seconds, or one quadrillionth of a second—the shortest interval currently in our control) and reassembled in step without transit delay time, it would take about 10 trillion seconds (10 to the thirteenth), or about 300,000 years to do the job. That is enough time for a lot of channel surfing, or death by overadvertisement if it happens on a *Star Trek* episode. If the disassembler works simultaneously on the whole surface of an object rather than one atom at a time, the task can be whipped through in only a millisecond (1/1000 of a second). But who has a multichannel disassembler, much less a single channel one?

The order of disassembly and reassembly also deserves consideration.

The *Star Trek* transporter always seems to remove and deliver people and things in a macroscopically random way, so that everything seems to turn into a mist or materialize from one. This presents a problem, especially for the early stage of dis- and re-assembly. In those cases it seems likely atoms would not enter a coherent arrangement, and gaps left at early stages would tend to collapse, perhaps fatally compromising the result. A better method would be to disassemble by removing atoms from the outside in, and reassemble them in reverse order.

Transporting people (or any living thing) atom-by-atom is an even worse problem. First, all life functions must be preserved at both ends. If the process takes 300,000 years, the traveler would not survive. The time would have to be cut to under a few seconds to prevent strokes. Second, the information requirement goes up because there must be great precision in removing and reconnecting all interconnections in the brain. This requires not just the right hookups but also the correct conductivity/capacitance at each connection. This is because memory resides in both the connections and their electrical strengths. It would not do to reassemble a normal looking person with a functioning body whose memory, speech, personality, and learning are all gone.

The times that I've calculated and quoted are based on a classical physics handling of the atomistic disassembly process. Without going into details, it is fair to say that by its nature quantum mechanical disassembly could be based on determining a group or combination state and thus could be faster by many factors of ten than the steps I outlined above. This has been demonstrated recently in elementary quantum computers. That would surely help, but such devices are much further down humanity's timeline.

SENDING ONLY EXACT INSTRUCTIONS FOR ASSEMBLY AND THE CLASS OF DISTANT CLONES THAT WOULD RESULT

There is another possibility to consider. If the complete body configuration of the transportee could be coded, then the *information* could be sent, instead of any matter, to an assembler at the destination. If all went well, the traveler would rematerialize and step out of the assembler at the other end. Presumably this information would travel as light or radio waves, and the speed of transmission would be the speed of light.

One must ask what would happen if errors occur during transmission: The encoder could make an error. The transmitter could make an error. The signal could well be degraded by its passage through the wispy gas clouds of interstellar space. It could be affected by its passage through planetary atmospheres. The assembler could make errors, too. In fact, on general principles, one must expect a non-zero error rate no matter what. It is possible that the transmission could include error-correcting codes to deal with the problem, but some level of information loss would almost certainly persist.

This leads to the question, what error rate would create a non-functioning traveler at the destination? At present there is no way to know. We don't have any quantitative data about what damage would be done to an individual due to the disappearance of brain cells or heart cells or the cones and rods of the eye. Such knowledge would be useful for medical purposes now, but its acquisition is many years away.

This method of transport suffers from another problem. Every time an individual is transported a new copy of him or her is created because the original is only measured, not disassembled. Therefore, with this technique a person could find herself in a universe with many identical copies.

This might have its uses. A diplomat could be sent to multiple star systems simultaneously to deal with an interstellar crisis and the home world would benefit from having one master diplomat deal with each situation. Of course if these copies lived for a while on their different worlds they would gradually become less identical, as each suffered his own special "slings and arrows." If travel by some other means could be arranged they might actually meet each other. If not, each would lead a separate life and die a different death. Keeping track of all the copies might get confusing, or it might force them to take extra names such as Erica 2255-1, Erica 2255-2, etc. It is even conceivable that the copies would be killed off after serving their immediate purpose, just to avoid such confusion, though I think ethics should prevent it. If the clones survived, widespread practice of this transport method would add to population problems. And such a system, because the original remains behind, could not be used to solve crowding problems on the home world by exporting the population.

One could argue that this form of mass transport might be involved in the abductions because some abductees claim to have come face to face with themselves while on board alien ships. But where is the scanner that inputs the data? And why would some aliens do this and some not?

QUANTUM ENTANGLEMENT AND TELEPORTATION

There is a surprising alternative to interstellar travel through wormholes that has recently begun to be examined, and it is related to the problem of matter transport. It is puzzling and counter-intuitive. This, of course, is because it is based on quantum theory. The phenomenon is called *quantum entanglement.* It, like much else predicted by quantum theory, flows from the Heisenberg Uncertainty Principle, which states that it is impossible even in theory to measure at one time with perfect exactitude certain pairs of numbers about the physical state of any piece of matter. For example, one such pair is the position and momentum of something, and another is the energy and lifetime of something. This is not a matter of failure in practice to have good enough equipment or skill. It is that every means of measurement requires the expenditure of energy and momentum and these expenditures affect the state of the thing being measured. We discussed the uncertainty principle earlier. Here it applies to the relations between two pieces of matter or two photons traveling separately from a common origin.

If we prepare two photons from one light source with the same state of polarization, so that polarizing sunglasses would let them both through, then they are predicted by quantum mechanics to share a combined physical state and behave in certain respects after that as if they were still together in their place of origin. This behavior is predicted to persist even if the photons are diverted by a beam-splitting mirror into two beams that travel in two different directions and over an arbitrarily long distance. If, after splitting up, one photon's state is changed, say by passing it through a crystal that changes its polarization, the twin photon's polarization will change by the same amount without any other intervention. Amazingly, this will happen *instantaneously* no matter how far apart the photons are when the first one is manipulated. At first blush it would seem that something wonderful could be made of this phenomenon—instantaneous communication leaps to mind, for example.

Unfortunately, for reasons to be explained, relativity must be served as well as quantum mechanics, and instant messaging appears to be impossible. However, a form of matter transmission may be possible.

It is completely routine in physics, and especially optics, to prepare beams of light or streams of photons from various light sources, especially such pure sources as lasers, using partially and completely reflecting mirrors, and to combine them, split them and create wave interference among them. It is also routine to measure their polarization using combinations of polarizing filters and detectors, such as TV cameras or photocells.

Using such tools, let us prepare a pair of polarized photons—"Jane" and "Joe"—from one light source, and trap them in perfect but separate containers, isolated from outside influences. They are entangled and their polarizations are identical, though as yet unknown to us because we have not yet measured that property. (If outside influences are allowed to intrude, the photons lose their special relation and become disentangled.) For the moment, they are the "perfect couple." We hand the "Jane" container to our astronauts, along with polarizing filters and detectors, and all are transported into Earth orbit, and then beyond, out to orbit around Mars. (Note that this voyage takes about eight months.) We maintain "Joe" here. After the astronauts arrive in orbit, we communicate normally by radio, which travels at light speed. This involves a travel time varying between about 12 and 24 minutes, depending on where in their orbits Earth and Mars are at the time.

In an emergency, it would be useful to have instant communication, but our Jane and Joe setup can't provide that. The reason is that for the simple setup described, using just two entangled photons, we have no way of sending a message. To do that requires measuring a change of state of the photons, in this case the polarization. To create a binary code, vertical polarization could indicate a "1" and horizontal a "0." However, to measure any *change* requires us to have knowledge of the original state of at least one of the photons. The problem with doing that is that to measure that state requires detection of the photon, and that requires its absorption by a detector, such as a TV. Thus Jane or Joe photon must "die," and the other is no longer entangled with anything, and no further change of state can be monitored. Furthermore, the result of the measurement can't be transmitted to the spaceship except by using other photons, which, of course, travel "only" at the speed of light. *Hmmm.*

A clever way has been found to get around part of the problem. It adds a third photon to Jane and Joe and some extra measurements. Call the newcomer "Mercury," the messenger, because Mercury will be "teleported" to carry the message. We send Mercury into the same beam with Jane, and they are made to form a sort of combined state with a joint polarization, which we measure. This measurement does not reveal their individual polarizations, only the combination. This combination measurement makes Joe instantaneously fall into the same joint polarization state, even though Joe is in the ship light minutes from home. There is even a bit of further news: Joe is actually changed to be in the state that Mercury had when it joined Jane, or in one of just three other states that are directly related. The problem is that the astronauts in the ship need to know exactly which of the four states it was in at the start. If they can determine that, then, in effect, Mercury has arrived on the ship, in all his glory.

By adding one more photon and one more detector, we can look for Jane. When she is measured to have arrived in one of the detectors and in one of the polarization states, all the information needed to make Joe on the ship instantaneously into Mercury is satisfied. But the ship won't know about this until the result of our measurement arrives minutes later by radio, at the speed of light. That does not make a very interesting communication system, as I said above.

However, quantum mechanics long ago predicted that atoms and molecules can interfere with one another and become entangled, just as photons do. Quite recently and for the first time, such effects were demonstrated in the lab for small groups of atoms. It is anticipated that molecular entanglement will soon be demonstrated. We can thus anticipate transferring all the above remarks and techniques for photons over to atoms and molecules, in other words, to matter. That means that we would have a way to transport (as in "beam up") matter to some place quite distant *at the speed of light.* It is not faster than light, and it is not a stargate, but still it is faster than any sublight ship and would use a lot less energy.

There are some interesting consequences. Foremost is that messenger

Mercury's state on Earth is destroyed in the process, and it reappears in the form of a changed Joe out in space. That means only one Mercury ever exists, and thus quantum teleportation will not lead to a lot of stray copies (read "clones") of atoms, phasers, or people wandering around the universe.

Of course, as yet we have no idea how to prepare a single whole Jane person as a whole-person state, nor how to add a second person, a Mercury, in such a way as to form a combination state, or to detect that state, or to create and keep isolated an entangled assembly of atoms that make up a Joe, ready to be converted to Mercury when the time comes. And when the time does come, poor Joe is gone.

Insufficient isolation degrades and disentangles Jane from Joe, and after enough of this, they are no longer connected and the entanglement teleportation will fail. To make this all work, we may need alien engineering help.

I WILL THINK MYSELF TO YOU! MENTAL TELEPORTATION

Science fiction has long toyed with teleportation by sheer power of mind. All one does to travel anywhere is to envision the destination, and *poof!* one arrives. Usually this is portrayed as an instantaneous process, with no muss, no fuss. That is, no pain, no reassembly errors. In many cases, the effort is said to take mental and physical energy, and doing a lot of it is exhausting. One of the most exciting examples of a story with this ability at its heart is the classic tale by Alfred Bester, *The Stars My Destination*. In it the main character has the hidden potential to jump from place to place (to "jaunt"), which he only discovers when he is trapped in a life-threatening situation with no other means of escape. But escape he does. He eventually is also able to teach others to jaunt.

This is an incredibly seductive idea. Unfortunately, of course, we have absolutely no clue as to how to accomplish jaunting. No one has ever made such a claim. There have been claims by people that they can move objects purely by mental command (telekinesis). This is a "Jedi mind discipline" *a la* Yoda and Luke Skywalker. In the real world, Uri Geller's spoon bending is perhaps the most notorious "demonstration" of this. However, several magicians travel around in his wake pointing out that Geller does this by traditional, if clever, sleight of hand. Even if one suspended disbelief and took

Geller at face value, he does not teach or claim to be able to teach his technique. There is no model for this supposed telekinesis, much less for self-teleportation.

Assume for the moment that jaunting is possible. There are still a few constraints in the real world to be taken into account. First, you have to be able to envision your destination in some detail. This was taken in the story to mean that you had to have actually gone there at least once. If all other forms of travel are sub-light speed, then obviously at least one very long trip must be made to a destination solar system before you can jaunt there instantly. This difficulty was resolved in the story by the existence of interstellar warp ships, which could travel at multiples of the speed of light, so that any interstellar trip would be only months, not centuries or millennia. The character left home in a ship and jaunted home from it. For shorter trips, such as across a country, one would make the trip first in an ordinary way and then jaunt home.

A second problem, largely glossed over, is that if you transport yourself from A to B some energy must be expended. Because there is no model for jaunting or jaunting efficiency, there is no real way to guess at the energy requirement. If it were the energy needed to lift yourself to planetary escape velocity then you would need the energy equivalent to that in a conventional rocket. Where would it come from? Even if the energy were only the amount needed to push you along in a wheel chair, the same question applies. You might easily exert enough energy with your own muscles, but with your mind? Even that small expenditure would likely fry your brain.

Since this is a bulk transport scenario, your departure would generate a small thunderclap, as outlined above. It would not be a totally silent process. As to the other end, you would want to have a firm grip on the visualization of your destination or you might end up materializing inside of solid or liquid matter, or you might fail to rematerialize at all and wander forever in an unlimited mindscape.

Finally, one has to say that so far no abductees have claimed to be mind teleported or to have jaunted themselves.

SENDING THOUGHTS MENSA-A-MENSA

If sending ourselves across the universe by mental teleportation does not pan out, perhaps there is still hope that we can communicate mind-to-mind. This has been a standard hope for hundreds of years and the subject of various kinds of magic tricks for more than 150 years. The majority of people who claim to have been abducted by aliens claim that they communicate with them via telepathy. In some cases, this takes place while the alien stares into the abductee's eyes, in others the alien touches the abductee and then thoughts flow. The abductees sometimes describe the experience of telepathy as exchange of words (in the abductee's language), sometimes as a flow of mental pictures, and sometimes as a stream of pure thought, though what that means is not very clear.

The scientific world has been interested in telepathy off and on. Alfred Wallace, Darwin's contemporary and independent co-discoverer of evolution, was a spiritualist. Oliver Lodge, an Oxford physicist and active spiritualist in the 1860s, was interested in thought transfer as a way to communicate with souls after their death. The word "telepathy" was invented by Frederick Myers in 1882 in an article he published in *The Journal of Psychical Research*. A number of physicists, including apparently Pierre Curie, were interested in spiritualism around 1900.

Over the years, certain individuals have claimed to be able to read the minds of others. There are many anecdotes of twins, especially identical twins, who have claimed to know each other's thoughts without talking to each other. My wife has on occasion said that she has felt something like that with her twin brother. It might be true—or it may be that twins really think alike, or it may be communication by nonverbal and subliminal cues, or it may be coincidence.

In the 1930s, Professor. J. B. Rhine, a biologist at Duke University, became interested in telepathy and conducted experiments in which a participant would sit in a closed room and stare at one of four distinctive illustrated cards while at the same time another participant in a different room tried to visualize what the first guy was looking at. Rhine claimed results for some people that were better than chance. He supposedly saw the same results over varying distances with the same level of success and no measurable time delays. This led to widespread belief, wishful thinking really, that telepathy would allow instantaneous communication across unlimited distances. Later attempts by others failed to confirm Rhine's results.

A small research group at the Rand Corporation, led by Harold Puthoff, attempted remote viewing and telepathy experiments in the late 1960s. They, too, claimed better-than-chance results, but, again, these were not confirmed by others. There are various organizations still in existence devoted to exploration of this controversial area.

If telepathy does exist, what can explain it? The usual explanation seen in the scientific literature and in some science fiction is that thought must be broadcast directly from brain to brain by weak electromagnetic waves. The reason this is a common explanation is that it is well established that the nervous system runs on tiny electric currents. Varying currents radiate electromagnetic waves, like radio and microwaves. These weak currents are routinely detected for medical diagnostic purposes and recorded as electroencephalograms (EEGs).

In one practical area, brain waves are well known and used beneficially. That is in the field of biofeedback. Electrodes placed on the skin around the head can easily monitor the state of a person's brain waves. In its most usual form, used for self-training to meditate, the biofeedback circuit responds with an audible tone when it picks up alpha waves, typically present in states of sleep and meditation.

You would think with a technology that can pick up a 1-watt transmitter from the edge of the solar system that someone with a sensitive receiver would have detected broadcast thoughts already, if they existed; but so far no one has. Perhaps the signal is so unusual that it looks like noise, and we have simply overlooked it. Perhaps thoughts are radiated on a completely different energy spectrum, one that has yet to be discovered.

No matter how it works, if it were true that the speed of thought were infinite and the strength of thought independent of distance, thought waves would violate the laws of all known forms of radiant energy. The second violation is the more subtle case. All known forces (except the force between the quarks) act the same way. A center of force acts like an antenna. The waves or particles which carry the force are emitted and spread out in space. Because they spread out more and more with distance, like ripples spread when you throw a rock in a pond, their energy spreads out more and more and they get weaker and weaker. The rate of weakening is generally proportional to the distance they have traveled from their source, squared. Doubling their travel knocks down their intensity by a factor of four, tripling it by a factor of nine, etc. Any reported telepathy that violates this rule will automatically receive a wave of skepticism from the scientific and engineering communities.

If telepathy exists then it is rare indeed among humans. If aliens come to it naturally or have mastered it, that still does not guarantee that they will be able to send and receive thoughts with us, unless somehow we are all on the same wavelength, or they have developed machines to help out. For us to develop telepathy, we would need to start with at least one pair of people who demonstrably send and receive thoughts. Then we could perhaps find a way to detect thoughts with instruments, which in turn could allow us to detect thoughts from elsewhere and to train ourselves to broadcast or narrowcast. Loss of privacy could take on a whole new meaning. We might be open not only to aliens, but to our neighbors!

6

YOUR UNIVERSE OR MINE?

With one being on either side of her Jerry said they floated her "out the window, like the wall. It's like it's not there." She got "that paralyzed feeling again" as she was pulled up to a large craft.

(From *Abduction,* by John Mack)

TRAVELS BETWEEN DIMENSIONS

In some UFO reports, alien visitors are said to appear out of nowhere and leave the same way. In many accounts of alien abduction, such as the one quoted above, the aliens and often the abductees walk through or pass through walls with perfect ease. Because no "gate" or process involving visible expenditure of energy is associated with these phenomena, wormholes seem an unlikely explanation for them. Gates require the bending and rending of space at enormous energy cost, and they would almost certainly cause a visible discontinuity in local space allowing a view into a tunnel or to another place entirely (see Chapter 4). Transporters seem to be ruled out because the hallmark of transporters is that they cause things to fade away or disassemble in place (see Chapter 5). That is not what we are talking about here. The first thing that comes to mind to explain such things is the natural motion of a person or thing through the space it exists in, if the space has more than three dimensions. At first blush, at least, there is no necessity to tear, bend, or compress space to move through it. To move we must expend energy, the power of our muscles derived from food. But we move through our space naturally.

There is a second, perhaps more subtle, possibility that makes use of the phenomenon that is at the heart of all semiconductor chip-based devices. It is called tunneling and is related to properties of electrons predicted by the modern part of physics called quantum mechanics.

Before looking at the possibilities of tourists from and touring to non-3-D spaces, and at the utility of mass quantum tunneling, we should consider why we can't normally put our hand through a table or walk through walls in our own space.

TOUCHING NOTHING: THE EMPTINESS OF SOLID MATTER

You probably remember from your school years the model of atoms that pictures them as tiny solar systems, with the nucleus of protons and neutrons at the center and electrons in orbit around them. The nucleons each carry 2000 times more mass than each electron. The protons and the electrons carry electric charge of the same magnitude but opposite sign (protons +1, electrons −1).

You may not have thought about how a scale model of this picture would look. If drawn to scale starting with the overall size of the atom, the nucleus and electrons would take up less than 1⁄10,000 the distance between them. If an atom were drawn to be 30 feet in radius, the electrons and protons would be less than three one-hundredths of an inch in diameter, a bit larger than the period at the end of this sentence. So it is really a fair statement to say that atoms are mostly empty space. This is true even though the more modern idea of atoms has the electrons smeared out in strange-shaped lobed clouds, instead of as orbiting tiny particles.

We solid beings can easily pass through a gas, like the air. In the air, the space *between* atoms is substantial, and they are not bound to each other. In the absence of a breeze, they bombard us from all sides, but we can't feel their individual impacts, we can feel it when their average speed goes up and the air feels warmer, and if it goes down the air feels colder. If we jump into a liquid such as water, we can pass through the liquid, but only in the sense of pushing it aside. The molecules of water are closer together than in a gas, and collectively they have a shape that depends on their container. Where the water meets air, it forms a surface. In neither case above do the *interiors* of atoms and molecules, which we noted above are mostly empty space, pass through one another like we walk through a gas. We pass *between* atoms and molecules when we do that.

So why *don't* atoms simply pass through one another's interiors when they approach one another closely? And, on the other hand, why do they stick together to form molecules, given that electrons, being of like charge, repel one another? To answer the first part of the question, we remember the following fact When "far apart," the electrons in different atoms (being of the same electric charge and being on the outside of the atoms) repel one another strongly (much more strongly than their individual strengths of gravity attract them, for example). Therefore, they bounce off one another. The repulsion grows as the distance between them shrinks (that is the nature of the electric force), and they would not stick together at all to make molecules except for two things: First, the electrons all have spins. If the spins of the electrons in two different atoms point in opposite directions they slightly attract one another and can approach more closely. That brings them close enough so the electrons of one atom can feel the attraction of the oppositely charged protons in the nucleus of the other atom. This can allow the atoms to stick together and become a molecule. However, the electrons of one atom can't pass into and through the electron clouds of the other atom because though they may "feel" their neighbor's protons and are in their grip, they are still repelled by their neighbor's electrons. They are caught between a rock and a hard place until some addition of energy breaks the molecule apart into its constituent atoms.

The bottom line is that no known technique will allow us to walk through walls the way we can walk through the air. There are, however, as mentioned above, two main ways that might be exploited to allow this to happen, or at least something similar.

CONTROLLING THE CLOUDS OF ELECTRONS; QUANTUM TUNNELING

At the beginning of the chapter, the phenomenon of quantum tunneling was mentioned as being important to the operation of silicon chips and possibly to walking through walls. But what is this tunneling? As it happens, it is probably the single most significant and yet unrecognized manifestation of the arcane physics of quantum mechanics in everyday modern life. It is all around us in its effects, but its nature is almost unknown to the general public.

The simplest way to understand the phenomenon of tunneling is to examine the operation of transistors, the electronic circuit devices that are at the heart of the chips and whose invention in 1947 sparked the chip revolution. Functionally, transistors replaced vacuum tubes as the switches and amplifiers of choice, and because they were so small and parsimonious of energy, they led immediately to miniaturization of all control, amplifier, and computing devices. Advances in theory and practice have reduced the size and upped the speed of chips by a factor of ten or more every 10 years. This trend is continuing, and there is now conceptual design under way for molecular-based computers.

Any material that is not a good conductor, such as a metal, and not an insulator, such as rubber or plastic or wood or ceramic, is what is called a semiconductor and has properties between those of conductors and insulators. There are a number of such materials, but the one in the most widespread use is the element silicon (not to be confused with *silicone*, the rubber compound used in adhesives and breast implants). Silicon is not in short supply, because it is part of almost every grain of sand and most rocks on Earth. By far, the most chips for computers and Walkmen are silicon-based, and the element lends its name to Silicon Valley, the new nickname for the Santa Clara Valley in California.

Transistors typically are a sandwich of two types of silicon compounds. One has an excess of mobile electrons inside, the other a deficit. To achieve these differences, the silicon is mixed ("doped") as it is grown into crystals with traces of other elements that spread evenly throughout the silicon. These p-type and n-type silicon materials are cut into thin slices and mated together to make transistors. The difference in electron mobility sets up a small electrical pressure between the layers, and if a small additional voltage is applied across the interface, the electrons will flow in relatively large quantity.

However, all is not quite as it seems. The interface between the layers actually sets up a mild electrical barrier between them. This can be thought of as an electrical "mountain." It is sufficiently "high" that the small electric voltage mentioned above would not actually push the naturally available electrons over the top to start the current and keep it going, according to Newton and Maxwell, the bright lights of classical physics. But for the quantum effect called tunneling, the transistors and chips would not do their tricks. (This effect turns out to be important in the Sun, too: without it, the Sun would not fuse hydrogen into helium at the same intensity it does now. Woe for life as we know it!)

At the turn of the century, discoveries of experimental physics required a revision of our understanding of matter and energy. It was found that light, which is energy, acts in some cases like waves and in others like particles. Then it was predicted by quantum theory, which arose to put a mathematical underpinning to these phenomena, that matter would show the same behavior under certain circumstances. It was predicted that a prime example would be the behavior of electrons when confined to atoms or to tiny containers in general. The prediction was confirmed by an experiment that showed that electrons traveling through the tiny molecular confines of crystal layers are able to be diffracted into beautiful patterns, such as light traveling through a tiny pinhole or water waves traveling through a narrow opening, all wave phenomena.

When electrons are confined to a tiny sheet of space right next to one side of the interface in a transistor, they tend to act like waves, which are spread out in space. Parts of the electron-wave have a finite probability of being on both sides of the barrier, and there is a real chance that the electrons can appear on the far side, even though the voltage applied does not give them enough energy to do that according to the classical particle-matter idea of the electron. Because we can picture the chip interface as an energy mountain, and the electron "burrows" through it rather than rolling over it, this process came to be called quantum mechanical tunneling.

So the tunneling effect permits transistors to work and millions of tiny transistors to be imbedded in a chip to make megahertz-speed (now gigahertz-speed) computer chips. The electrons can tunnel through, but does this mean whole atoms can do the same thing?

"FALL IN, YOU ATOMIC SHIRKERS!": MAKING ORDER OUT OF CHAOS

The fact that chips make their currents based on tunneling may suggest that we are just a step away from being able to tunnel our way through walls. Unfortunately (or fortunately, considering our desire for personal privacy), that is not true. While it is the case that any ordinary electrical current consists of a stream of many electrons, and they flow in the same general direction and thus have a sort of order, the order is very limited. They actually flow in an individually uncoordinated way, bouncing off one another in randomly varying directions with randomly changing speeds. The current represents their average flow.

To walk through a wall successfully, we would need to preserve exactly our body's preexisting pattern in space, our total atomic and molecular construction, as we passed through it. If a significant percentage, say, 10 percent of the molecules in your brain's neurons are rearranged, there is a high likelihood you would lose part of your memories and also part of your brain's functions. Perfect order in the tunneling, as mentioned above, is not what normally happens. However, quantum mechanics predictions are based on probabilities. Even though the idea seems contrary to common sense, there is a calculable probability that just by chance all of the atoms of your body could pass through a wall. The probability of that happening compared to transistor tunneling depends on the number of atoms and molecules in your body compared to the electron-by-electron probability in the current (the more particles that must perform together, the lower the probability) and on the wall thickness compared to the interface thickness in the transistor (the thicker the barrier, the lower the chance of tunneling). Given the numbers involved, the event might indeed happen by chance, but not unless you wait longer than the lifetime of the universe! This is an interesting and natural way to walk through walls, but it is not a useful one.

Must we give up on tunneling? Not necessarily. There might be a way to use artificial tunneling. One of the more interesting aspects of quantum mechanics, mentioned above, is that it predicts that in certain circumstances matter and energy behave alike, especially if the matter is very small, such as electrons, and confined to small spaces, such as atoms. It is appropriate, therefore, to ask, could whole atoms or molecules (which are very small though not as small as electrons) be made to act like electrons? And could we find a way to take control of this behavior so that is completely orderly and not random?

The answer to the first question has been thought to be yes for decades, because the equations of quantum mechanics predict it, and other bizarre quantum behavior has been observed of atoms and molecules. But the possibility of wave-like behavior of whole atoms and molecules was not taken seriously until recently. Just in the last three years, experiments have demonstrated that whole atoms can act as waves and interfere with one another like light or water waves. To do this they must in effect pass through one another, just as we have been hoping they would do. The experimenters have prepared atoms to travel in a relatively orderly fashion, so that they behave as if they are waves whose crests and troughs have a definite relationship. The result is that as they pass through one another in space, they add or subtract from one another in a pattern of ripples, like waves in a swimming pool striking a wall and bouncing back on themselves. The atomic pattern is detected using atom-counting detectors and can be discerned only after many atoms have passed through the apparatus. The pattern only occurs if the atoms have a fixed relationship between their "crests" and "troughs." (Yes, it is weird to speak of atoms as having wave properties, but that is a main burden of quantum theory and of the world we live in, and we need this behavior if we want to walk through walls.) These atoms, by virtue of this orderliness, are said to be *coherent* with one another.

Is this experiment with coherent atoms sufficient to demonstrate that walking through walls is in our future? Not exactly. The current efforts control perhaps thousands of atoms, but a typical human (alien) body contains perhaps 100 billion billion billion atoms. There is no known way to scale up to that requirement. Here we might look to lasers for a useful analogy. The reason is that lasers work by doing something somewhat similar.

Inside lasers, atoms of a preselected element are induced to exist in just one electron energy state or orbit. They are chosen to have a particular orbit-to-orbit electron energy jump that is at a useful color. They are enclosed between mirrors coated to reflect just that one color, spaced apart at a distance that is an odd or even number of wavelengths of that color. In other words, they are enclosed in what is called a resonant cavity that has a natural frequency corresponding to that color, like an organ pipe resonates for one pitch. These laser atoms are forced to emit light at that one color, whose waves travel precisely in the same direction and are all in step trough to trough. The light of a laser is thus said to be coherent for the same reason as for what might be called the "atomic laser" described above. The atoms in a laser are forced to emit coherently by the strength of the laser light inside the laser.

Perhaps this provides us our clue. If we could find a way to create a strong atomic beam control, perhaps inside some sort of resonant cavity, we might induce a whole body's worth of atoms to behave coherently. But we must face the fact that the body contains all sorts of atoms and molecules, a number that is not entirely cataloged but perhaps in the thousands, so it would seem we would need to find and control the resonances for all of these types. While this is not ruled out by known physics, there is no road map to that technology.

A very limited control is exerted over the body's atoms and molecules, *en masse*, when we undergo an MRI session to image the soft tissues of our body. The strong magnetic field of the scanner combined with a radio signal makes everything sit up and take notice. But it can't enforce the order we need or detect the state of individual atoms so that they can pass through one another. If it could then almost certainly accidents would have happened. So far, no one has spontaneously welded to or fallen through any MRI table.

It might be that, in the future, some large piece of equipment, based on the current experiments, might eventually be able to make all of the atoms and molecules in a body coherent. But if that is what is required then it might be simpler just to open the door and walk through normally. Of course, an advanced alien civilization might, with its long head start on us, have devised a way to miniaturize such a device, and that may be what allows them to walk through walls and drag humans with them.

MIND OVER MATTER?

Thinking about miniaturization of the matter control device suggests another question: is there any way that a thinking being might substitute itself for the controller by use of pure thought? At first blush there might seem to be no way to do that, but perhaps there might just be one. It would have to be an extension of a mental technique of self control to an extreme degree. There are various mental disciplines that humans have developed over the years to gain control of their bodies. Traditional techniques include self hypnosis and perhaps principally Yoga. They have allowed people to learn to control heart rate, breathing, and, in some cases, blood flow and pain. Some of these functions, especially the latter, are outside the control of untrained people. However, these techniques would be difficult to apply to learning to control the states of atoms because there is no known natural sensation corresponding to atomic states that can be felt by the brain.

This would be the end of the story except for the fact that in the 1970s a scientist conceived of the idea of connecting the brain to the body through an instrument with sensors that allow the detection of body states we don't normally feel. This technique came to be known as biofeedback. It was first applied to achieving a meditative state called the alpha state, one of the states achieved by Yoga practitioners and one known to coincide with the brain's generation of electrical signals of a particular form called alpha waves. These waves are also observed in one of the stages of ordinary human dreaming.

This was seen as a revolutionary development at the time. Among other things, it is a way of training ordinary people to control various usually uncontrollable bodily functions using straightforward feedback conditioning in a relatively short period of time. Mastery of Yoga takes years. So perhaps there could be a way to apply this technique, using the right sensor to detect the state of your atoms, to control them. However, the devil resides in the details of these last two tasks.

First, though we have detection schemes that can measure the states of atoms in the way we would like for our purpose (walking through walls), they require the lab setup and large equipment as described above. Furthermore, we can only deal with hundreds or thousands of specially prepared atoms, and as mentioned above we would need to put order into the ordinary unprepared body atoms and molecules, in all their variety and gigantic quantities (10^{29}—10 to the 29th power!), all dancing to different drummers. There is also the problem that we almost certainly have no natural signal to fasten on and use for feedback, as we do for alpha waves. We would need to invent one, and then feed it to our brain. This may not really be a major problem because we already do something similar for conventional biofeedback. No one thought there was a direct way to detect alpha waves either. The biofeedback technique converts them to an audible tone. After sufficient training the practitioners can throw away their equipment and do without the tone while achieving the alpha state. They can feel it internally in some way.

This raises the next problem: We will be trying to achieve a state of being that is almost certainly entirely new to humanity. Doing this for the first time might prove extremely difficult because there will be no reference state known from common experience. It may thus also be impossible to achieve continued success without the sensors and other equipment to tell us about our internal order.

We could probably accept the need for big equipment and a non-portable lab setup for doing the training to master this atomic control, if it later freed us to do the control outside the lab without any assistance. And if somehow that turned out to be possible, we would still have to contend with the other side of the equation: we have to detect the state of the atoms and molecules of the wall in order to mesh with and pass through them. So a second detector/controller would probably still be needed.

Learning to mentally detect the internal state of something external to us seems even more unlikely than the already unlikely possibility of micro-controlling one's own state. Doing so would require us to acquire what is in effect a new sense. The sense would have to be capable of detecting the state of atoms and molecules inside something else as well as on the surface. We have no such sense now. You could suggest that sound might work because it penetrates most materials quite well and dolphins' refined sonic capabilities allow them to "hear/see" into things, according to some experimenters. However, the sound required to detect individual atoms and molecules would be so short wave that we know of no natural or artificial way of generating it. And because sound is not really electrical or magnetic in nature, it can't be used to impose detailed order of the required sort on atoms and molecules.

We would need a new organ or a way to use an existing organ to create this electromagnetic control. We might be stuck here, but that does not mean that some new sense might not be found. In any case, there is a completely different potential approach to our problem.

LOCKED ROOM MYSTERIES: ENTERING FROM A NEW DIRECTION

We have been discussing the severe problems that occur in the attempt to walk through walls. However, there might be a way to enter a space from outside that bypasses the walls altogether. This might seem like a nonsensical statement but in some ways it is easier to conceive of than imposing detailed molecular order. The idea that you can enter and leave a locked room without passing through the walls or an opening may be bizarre but it appears to be allowed by mathematical theory. Not only that, but it is not a new idea. It was discussed in great detail in a fantasy novella called *Flatland*, written by E.A. Abbott, first published in 1884. It is still available in an inexpensive paperback edition and well worth reading, though its style is rather dated.

Abbott conceptualized a world of two dimensions, which he called Flatland, peopled by various characters and castes that he created to discuss the properties of things in a 2-D space and thus to illuminate our unexamined notions and assumptions about our existence in a 3-D space. He also discussed 1-D beings. We will follow in his path to look for ways of understanding what would happen if beings normally living in a space with four or more dimensions tried to enter our space. We hope there are valuable analogies between "going down space" from three to two dimensions and "going down space" from four to three. This is not, however, guaranteed, for reasons having to do with the physics (as opposed to the mathematics) of space and matter. There are a number of recent books that center on or overlap this subject. They include *Hyperspace* by Michio Kaku, *Surfing Through Hyperspace* by Clifford Pickover, amusingly and interestingly illustrated by the author, and perhaps best of all, *The Fourth Dimension* by Rudi Rucker. They would all repay your interest.

What did Abbott depict? He described Flatlanders of simple geometrical shape, such as straight lines, triangles, squares, and polygons. The social status was determined by the number of their sides. Women are the straight lines, and the highest nobles have many sides. The middle-class men are squares or pentagons. The inhabitants with so many sides that they look like circles belong to the priestly class. He described pentagonal (and higher order polygonal) dwellings that had doors (but no windows). He assumed a thin haze permeates the atmosphere of Flatland, that rains come from the "North," and that every part of Flatland, both inside and out, is uniformly illuminated by an unknown source (which he later explained). He ascribed speech and motion (and emotion!) to the inhabitants. He gave them human motives and failings. He discussed how they would appear to one another. He also described what he thought Flatlanders would see if a person from our space (he called it Spaceland) entered theirs, and also what the 3-D visitor would see of Flatland and Flatlanders.

The presence of the haze in Flatland was key to Flatlanders being able to visually distinguish their fellow shapes from one another. It allowed the perception of varying distance to parts of a person, even when the viewer was face on to it. For example, when facing a triangle vertex-on, the places on the sides would seem fainter the farther from the point one looked. Haze and shading would give distance cues. The arc of a circle would also fade as one looked along it away from its closest point, but the fading would look different than for a triangle. A polygon would reveal itself by fading away in short segments between discontinuities. Failure to learn to visually recognize their fellows could lead to fatal accidents. Failing to notice the point of a triangle (soldier) or the tip of a woman could allow one to impale himself.

Abbott in his history of Flatland describes how color briefly entered as a fashion, which allowed additional cues, but was done away with when it was found that members of the lower castes falsely assumed the colors of the upper castes and thereby defrauded and dishonored women who were intent on marrying at or above their station.

Abbott created his Flatlanders with functioning eyes at their vertices. He omitted any details of their inner workings.

Abbott maintained that the inhabitants' exterior shapes were visible, but not their interior contents, unless the skin or outer boundary of the Flatland object was breached. He rather inconsistently (but this was fiction!) maintained that sound could be heard everywhere from both within and outside Flatland.

THE TRUE APPEARANCE OF HUMANS IN FLATLAND

Abbott made his 3-D visitor to Flatland a sphere, for simplicity. He then proceeded to show that Flatlanders, if they were present when the sphere arrived, would see an arc in their 2-D space, because there is no third (height) dimension. The arc, in the case of a visiting sphere, would actually be a circle in Flatland. The visitor would first appear as a point in Flatland, then as the sphere passed through the plain/plane of Flatland it would grow in length until the sphere's equator passed through, then it would shrink back to a point and disappear. The Flatlander would, if visually trained, recognize the visitor's arc as actually being a circle of varying diameter. If you are having trouble visualizing this, think of slicing a sphere into a set of parallel thin disks, and then seeing them only one at a time and edge-on, in succession.

Now that you have put yourself in the Flatland narrator's shoes what would you see if you lived in Flatland and were visited by a cube or a polygonal solid (one with many faces) passing through face-on to the space? The answer in the first case is a square, which would appear suddenly, be there for a while, and then disappear. In the second case, you would see a particular polygonal perimeter depending on the shape of the visiting solid. The number of polygon faces would probably vary during the passage. All this regularity assumes the passer-through is face-on to the space as it enters and leaves and does not rotate as it transits. If it were corner-on or did rotate, the succession of shapes would be more complicated.

What about the visit of a shorts- and tee-shirt-wearing, be-sandaled human, or for that matter, an octopus, to Flatland? In the former case, if our feet penetrated first and we sank through Flatland standing up the inhabitants would see a pair of shoe-sole perimeter shapes, then footprint-shapes, toes protruding (Flatlanders would be mystified because they are described as having no feet!), then hairy circular sections of our legs, then the cloth rings of the shorts, then the figure eight shape of the crotch and the oval shape of the waist, then the oval shape of the tee-shirt, ribs hidden, then the narrow circle of the neck, then the hairy surface of the neck and skull base, the bulge of the lips and then the nose and hair and ears, then the hair at the crown and then nothing. All seen edge-on. Bizarre!

And the octopus—what an entanglement! The view would depend on how it disported itself: tentacles trailing, curled or knotted, beak revealed or concealed, etc. Can you picture seeing the tentacles and suckers in section? What if it was moving in a lively way as it passed through? Then the Flatlander would see lots of moving almost-circles, floating back and forth in space in completely unpredictable ways.

THE APPEARANCE IN OUR SPACE OF A 4-SPACE OBJECT; CUBES AND HYPERCUBES, SPHERES AND HYPERSPHERES

All these visualizations, mental calisthenics, if you will, are to prepare you to think about what we would see if a 4-D being visited us. And though this might seem a modern idea, Abbott himself considered it and for a particular case gave an accurate depiction. He reasoned that there is actually a rule that relates the number of faces and sides of a regular shape in each level of dimensionality. He used the example of a line moved sideways parallel to itself to generate a square, and the square moved "up" parallel to itself, a direction a Flatlander could not recognize, to generate a cube. Then he reasoned that in a 4-D space the cube could be moved parallel to itself in a new direction, one we can't see, and generate a 4-D cube, what we now call a hypercube. It would have eight cube faces. He even suggested how to visualize this 4-D thing. Here we will depart from Abbott to describe several ways to think about the hypercube's appearance to us.

There are actually several ways to visualize a hypercube in 3-D. This should not surprise you if you remember that a 3-D cube can be visualized several ways in two dimensions. This is a way of saying we can draw a cube on a sheet of paper (which after all is almost 2-D) in various ways. Drawn face on, it would be a square. Visualized as opaque but seen from in front, above, and to the side it would be drawn as a set of squares and parallelograms with the rear lines hidden. Seen from in front but as a wire frame made of its edges, it would be drawn as nested squares, the smaller connected to the larger by straight lines at the corners. The wire frame version with equal-sized squares easily gives rise to the optical illusion that the front and rear faces can change places as you stare at the picture.

The case of the hypercube is analogous but, not surprisingly, more complex because more dimensions are involved. One visualization is that of two nested cubes whose corresponding corners are connected by straight lines. There are eight volumes enclosed in this representation, and to us they look like two cubes and six symmetrical truncated pyramids, all touching their nearest neighbors. In four dimensions, they should all be cubes, as mentioned above. This misrepresentation is the result of our depiction being one dimension down from their true space. (It is possible to produce a beautiful 3-D version of this visualization using soap bubbles. Make a cubic wire cage, edges only, with a handle wire attached to one corner. Submerge it in soap bubble soap and pull it out, and often an inner cube will form spontaneously.)

A different version pictures the hypercube as a wire frame drawn as if the cube had been repeatedly displaced diagonally from its original position. The result is a complicated looking drawing of an octagonal solid divided into parallelogram and rectangular solids. It is hard to see such a figure as a solid and yet prevent it from "flipping" its orientation. The third way to depict a hypercube is to show an image of eight cubes. The usual representation is of four cubes in a line, with four others anchored to the exposed faces of one of the four in the row not at either row end. In 1954 Salvador Dali used this version of an unfolded hypercube as the crucifix in his painting *Corpus Hypercubus*, and it can be viewed in stereo in some publications. For an amusing take on what would happen if you built a "real" hypercube in our space, read Robert Heinlein's 1940 classic sci-fi short story *And He Built a Crooked House*. And I should mention another sci-fi short story, Armin Deutsch's 1951 tale, *A Subway Named Mobius*, which deals with another version of the problem of the intersection of 3-D and 4-D space (see below).

What if the 4-D object is curved? What would we see if it passed through our space? This is a more complex thing to visualize, but for simplicity consider the analog to a circle in 2-D and the sphere in 3-D: a sphere in 4-D, which we call a hypersphere. In some sense it will be like a hypercube, that is a packed cluster of spheres, but it is hard to picture an arrangement that leaves no gaps in the 4-space, as you would see if you tried to stack a bunch of oranges together, for example. If the cluster passed through our space, we might see a set of spheres appear and disappear. A computer simulation, reported in *Fourfield* by Tony Robbin, showed that using a central projection the section we would see of a hypersphere would look like a torus (doughnut). As the sphere rotated in 4-space, we would see the torus become multiple linked toruses of varying cross section and orientation, shrink to columns and spread out into flat planes, and finally switch positions and appear to turn inside out.

INVADERS FROM OUTER 4-SPACE

It is easy to conclude from the foregoing that what we would see of a 4-D creature as it visited us would depend critically and in a complex way on its shape and behavior in four dimensions. If a creature with a large 4-D hand reached into our space, we would probably see a set of blobs appear suddenly from out of the air, coming from all directions (think of our hand reaching into Flatland to pluck a Flatlander out of his universe). Does this sort of thing correspond to encounter reports of the appearances of aliens? No. People typically see a range of alien types, including short grays and tall leaders wearing "normal" clothing, and they walk and talk as we do (though in some cases apparently use telepathy). This suggests that if these aliens are truly inhabitants of 4-space, then they must have a very special shape and act very circumspectly as they move in 4-D because they look and move normally in 3-D, except perhaps for walking through walls and rising effortlessly out of buildings and up into space ships.

How would this work? For their appearance to be normal, one possibility is that, for whatever reason, one "face" of their 4-D shape looks like a 3-D person. Another way of stating this is that from one point of view their projection into 3-space looks like we do, more or less. Perhaps they design their "contact" versions to look something like their contactees so they won't scare them unduly. By analogy, picture yourself putting your hand with a Flatlander shape painted on it, palm down, into or onto Flatland. Without much difficulty you could keep it flat and oriented as you please with respect to any Flatlander, always assuming you can actually see Flatland and Flatlanders as a 2-D sheet with 2-D shapes jostling around on/in it.

Another possibility is that one of their four dimensions is infinitely thin, so that they are actually three-dimensional, even in their own space. To hold their shape stable in our space, they would have to pay constant attention to their orientation. This might not be too difficult. The analogy would be if we could somehow make ourselves two-dimensional, looking like the Flatlander shapes of triangles, polygons, etc., and then lower ourselves flatly onto/into Flatland. We would have to be careful not to raise or lower ourselves or tip ourselves in the height (3rd-D) direction, or parts of us would disappear from Flatland. Back home in our space we might find that we would lead a difficult existence. How could we ever stand up? And going up-dimension how could a 3-D inhabitant of 4-space avoid similar problems? He would likely be unable to "stand up" in the sense of his height (4th-D) direction. It may be that this severe thinness is the biggest difficulty standing in the way of the existence of 4-D beings that would look like us when they visit our space. To visualize such thin creatures in our space, think of the toons in the movie *Roger Rabbit.*

Suppose 4-landers could enter our space. Would they be able to "walk through walls," the task we originally set? No, not exactly. Without the supreme subatomic control we discussed above, they could not do that. But they could do something that looks a lot like it. They could come from the fourth dimension and enter our space from the 4-direction just next to the wall. They would appear to come out of the wall, but what they would actually do is enter just in front of it. Assuming they could see the wall from their 4-space, they could enter as close to the wall as they like. To appear to walk through it, they would only have to walk "around" the wall by taking a detour in 4-space. To abduct a human and appear to take her through the wall, they would have to drag her out into 4-space, walk "around" the wall, and then return to our space just outside the wall.

This, I expect, sounds fairly miraculous. But even more amazing feats might be possible. If the 4-landers can see us at all, they will be able to see our insides as well as our outsides, because they have the freedom to look along a direction we don't have. This might seem perplexing but again think of Flatland. If we can see it and see into it at all, we can survey the innards of any Flatlander because we can see them from our directions "above" and "below." They, however, can't see into themselves. Similarly, a 4-lander should be able to see into us.

Some authors have suggested that this might enable a 4-lander to perform bloodless, painless surgery on us. I don't think the operation would be bloodless or painless, though it could be done without breaking the skin. All the 4-lander would have to do to is reach into the patient from its special 4-direction and it would avoid the skin just like it was walking "around" the wall as described above. If it could interact with us 3-landers at all then it would presumably be able to cut or manipulate as a normal surgeon would, but there would then be destruction to tissue and nerves and the resulting internal bleeding and pain. Anesthetic, it is true, could be administered without a needle injection because it could be poured in from the 4-direction. But the patient would still feel the hands of the surgeon if not under general anesthetic.

This sort of operation actually matches some abduction accounts rather well. Typically, when people are examined the aliens use surgical instruments, which enter the abductee bloodlessly and leave without blood, sutures, or scars (in some cases small scars are alleged to be evidence of the procedure), but the patient often feels pushed and pulled inside. Some women think that eggs have been removed, and they often feel a great deal of pain. If we assume that the instruments and examiner's hands move into the fourth dimension just at the skin and re-enter our space just inside the skin then the description fits the hypothesized 4-D procedure.

On a less medical note, a 4-lander could, it has been pointed out, lead quite a life of 3-D crime. By entering and leaving a bank vault in our space from a 4-direction in its space, all of the vault's external alarm systems would be avoided. Unless there were 4-D cops around, escape by the same route would be trivial. Likewise, the ability to travel in four dimensions would enable the traveler to save the lives of people trapped in 3-D buildings in case of fire or earthquake, or of miners trapped in mines. The barriers in our space would be circumvented (literally), and the distance to reach them in 4-space would be unrelated to and could be much shorter than in our space.

If we wanted to, could we trap a 4-lander, say that bank robber, in our space? Putting it in one of our ordinary jails would not suffice. It could simply step out along a 4-D direction we can't sense and disappear. If we can interact physically, then we probably can trap it, and no room would be needed to hold it. We would have to try to put a spear through any part of its body visible in our space. This might pin it to our space, like a pin through a corsage on the bodice of a dress.

MASTERS OF INTER-DIMENSIONAL JUMPS

There is yet another interdimensional travel possibility. There might be a way for 3-D beings to enter higher dimensions from our space and then return to it. They would be "masters of interdimensional travel" (or the old phrase, "masters of time and space" might apply). The problem is no current physics gives the slightest clue how to enter the fourth or any higher dimension. Space warps and stargates might manipulate our space and bend it or tear it, but that does not necessarily gain us entry into a higher dimension. As I stated earlier in this chapter, the only natural and easy way to hop dimensions is to go down-dimension. At least then we can see where we are going (maybe!). To go up dimension, we have to see or invent or find a fourth direction we have never experienced. In 4-space that direction would be perpendicular to the three directions we are used to. But in our 3-space, which direction is that? Try to point to it and you can easily become confused (and might be considered crazy, if you are not standing in a lab and writing equations on a blackboard). It is a direction literally out of our space. It is not obvious that any 3-D being would ever be able to visualize it, much less travel along it. This remains, for now, an open question.

CAN HIGHER-DIMENSIONAL BEINGS ENTER OUR SPACE? COULD WE KISS THEM?

Suppose that figuring out the right 4-D direction is not a problem. Is the path of interdimensional traveling paved with roses? I think it probably isn't because I believe the descriptions above are incomplete in areas related to aspects of known physics. When these are considered, the whole enterprise of interdimensional travel is cast into doubt.

Again let us begin with visits down-dimension to Flatland, if you will. First, is it physically realistic to think we can see or otherwise detect a 2-D space from outside? Of course we can look at flat sheets of paper and flat sheets of transparent plastic. But are these actually separate spaces from ours? Not really. For one thing, they are not truly 2-D. They might be thin in the third dimension but they are not infinitely thin. They are composed of many layers of atoms and molecules. Think thinner: picture a monomolecular film, such as a coating of gasoline on the surface of a puddle of water. In many cases it will naturally become one molecule thick. That is very thin, about ten one hundred-millionths of a centimeter (one one hundred-millionth of an inch), but it is not zero. An electron, though incredibly tiny, has been shown by experiments to have a diameter that is greater than zero. It is a tiny 3-D sphere. No known type of subatomic particle is thought to be 2-D. So we have to ask, is 2-D matter possible? I don't believe we know, but I would like to suggest that quantum mechanics, which we discussed when we were looking at passing through walls, might forbid it.

As discussed above, the theory includes the Heisenberg Uncertainty Principle. According to that principle, the greater the certainty in position of a particle, the more uncertain is its momentum. This is behind the phenomenon of quantum tunneling discussed earlier. It has also been interpreted to mean that the smaller the box in space an electron is confined to, the more momentum and energy it must have. This in turn leads to a way of explaining why electrons in atoms don't end up permanently bound to the protons in their nuclei. The nuclei are so small that the electrons must have a lot of kinetic energy when they are (briefly) localized to the protons, and they spend most of their time flying around in a cloud ten thousand times larger in radius.

Now extend this idea to 3-D matter in 2-D space. The uncertainty in the infinitely thin direction of a 2-D space would be infinitely large. So 3-D electrons might either rush around the 2-space very fast, too fast to bind to 3-D protons and make Flatland atoms, or they might simply radiate away to any 3-D space that might surround it, meaning that 2-D spaces would be empty of 3-D matter as we know it. But what about 2-D electrons and protons? Could they exist? I don't think so. That's because of the nature of electromagnetic waves, the type of energy that is fundamental to electron and proton properties. I think there are no Flatlanders.

So then one has to ask, what about 2-D energy? First, on a macro scale, is there any reason to think that Abbott was correct when he gave his Flatlanders eyes to see with? Could there be a functioning 2-D retina? I think not, because as I said above, I don't believe 2-D matter can exist, but also because I don't think 2-D electromagnetic energy can exist. When we speak of energy in our space, we speak of transmission of force by "virtual" particles or waves of energy. The particles include gravitons to transmit gravity, photons to transmit the electromagnetic force, and various mesons to transmit the strong, glue and weak nuclear forces. Gravity brings neutrons and protons together (with their accompanying electrons) in large collections of matter, such as the Earth and the Sun, while the other particles hold neutrons and protons together as particles, and to one another in atomic nuclei. These particles might be a sort of vibration in space, so we can ask, can 2-D versions exist? The objections above might apply.

The most important particles/waves for us, per se, are electromagnetic. The electric force binds electrons in atoms, atoms together in molecules, and molecules together into ordinary lumps of matter, such as us. Photons/waves transmit the spectrum that includes light and radio. In their wave incarnation they are thought to be 3-D, because they have a magnetic vibration and an electric vibration which are perpendicular to one another and to the direction of wave travel. (Picture two ropes seen from the side, one vibrating up and down, the other towards and away from you, as they travel together to the right or left.) For many practical radio problems we solve wave equations as if they were 2-D, so how could light propagate if there were not two dimensions for the vibrations as well as a third to travel in? For these reasons, I believe that 2-D electromagnetic waves and the static electric force do not exist in 2-space.

Of course the motivation for talking about 2-space is for thinking about 4-space and 4-landers visiting our space. Here the answer is even less simple. If a 4-D Heisenberg Uncertainty Principle applies to 4-D matter, it might suggest to a 4-lander (and to us) that 4-D matter can't exist in three dimensions. I conclude that it might be possible for our matter to exist in 4-space if and when we visit it (however, see below on string theory), but no true 4-D matter may be able to visit our space. Hmmm . . . So our visitors might be just what we said above, not from a 4-land, but from our universe, "masters of interdimensional travel," who can jump in and out of it at will. And if they can arrive at all, we should be able to kiss them.

COULD A BEING FROM A PARALLEL UNIVERSE ENTER OURS? WHAT IF THE LAWS OF PHYSICS WERE DIFFERENT IN THE OTHER UNIVERSE?

The arguments above lead me to suggest that in general it is unlikely that a being from another universe would be able to enter ours, if the other universe had more dimensions than ours. The kind of matter and energy simply might not fit into our space of 3-D. But what if there were other 3-D universes somehow adjacent to ours? If the problem of traveling from one to the other could be solved, perhaps through some sort of wormhole (see Chapter 4), then there is a chance travelers could enter our universe from another one. This would certainly be true if the laws of physics were the same in their universe as in ours. But how likely is that? And what if the laws were different?

At the moment, we don't have any way to know if other universes exist (see below), and if they do, we don't know if the laws of physics and all the important physical constants will be the same in them as in ours. Let's assume for the sake of the argument that they exist and that in some of them the laws of physics are different. A visitor from one of those tries to visit us. What would happen? There is no easy answer, if the great conservation laws, the conservation of matter/energy and the conservation of momentum, rule. But if the ratio of the electric and gravitational forces is different over there and if electrons and protons there have different masses from those in our air space, perhaps these parameters would adjust automatically when the visitor crosses over. But would that readjustment be benign? Or would the rate of the visitor's metabolism change abruptly, or the strength of its cells or skeleton change, or its size suddenly adjust?

If the conservation laws did not hold, or the force of gravity bent space in a different way over there, then all bets would be off. One might argue that the characteristics of the space you are in determine the properties of any matter in it (see below), and that, in effect, matter or energy entering our space would automatically behave "normally." But it is perhaps as likely that crossing over would actually cancel out the visitor.

TEN DIMENSIONS AND THE THEORY OF EVERYTHING: WHAT IF THE LAWS OF PHYSICS IN OUR UNIVERSE ARE DIFFERENT FROM WHAT WE THINK?

A number of theories of contemporary physics contemplate the possibility that there are many universes of which ours is only one. These ideas arise out of the territory where cosmology meets subatomic particle physics, and also out of quantum mechanics. These universes are thought to be seen as 3-D by their inhabitants (if there are any), like ours is. There is a school of quantum mechanical thought dating from the 1920s that interprets the Heisenberg uncertainty principle as saying that every observation of a physical thing directly affects its state. Thus observing a particular atom affects its position and momentum. Furthermore, quantum mechanics calculates each new predicted state as a probability, not a certainty (unlike the classical physics of Newton and, in this respect, Einstein). In this calculation it is assumed the most likely state is the one that results, but that all others have some chance of occurring. The reality of the other possibilities was disputed back in the 1920s, but one school of thought came down on the side of the idea that all these possibilities actually occur. The concept that follows from this is that new universes are being created moment by moment, resulting from the events of each instant, each a variant of the one we live in. These variants would proliferate from the actions of each one of us. The number of universes would be immense, if not infinite. No one ever found a road map for jumping from one universe to another under this interpretation. The quantum "many worlds" interpretation is now out of fashion, but the idea of multiple universes has reappeared as a result of cosmological speculations and the recent success of a major new theory of space, time, matter and energy: the string theory (and its latest wrinkle, the M theory).

In essence, this theory flows from the great insight of Einstein that matter and space are inseparable. But it also combines this insight with quantum mechanics to explain the existence of phenomena on the smallest scale. Einstein's theory of general relativity, his theory of matter and space, fails at the tiniest scales, but it actually falls out of string theory and, growing out of string theory, it works at the smallest scales. At least that is the way it looks. String theory is incredibly complex mathematically, and the results we are going to refer to are in some cases hard-won proofs, in others educated speculations based on insights into the way the theory works in the simplifying approximations that have been applied to it. The theory was first proposed in the 1970s and 80s, went into eclipse for a while due to its difficulty, but advances have brought it back with great progress in the late 1990s. The reason for discussing it is that it casts a new light on both our discussion of the fourth dimension and on the possibility of many universes.

The strings themselves were the first version of the *new* "most fundamental bits of matter and energy" and were originally thought of as infinitely thin tiny loops or line segments ("closed" or "open" strings) that vibrate at various frequencies (later versions under M theory include vibrating membranes and closed surfaces in three and higher dimensions). These strings are thought to be smaller than the smallest subatomic particles, and they are thought to be the actual things that we detect experimentally as nearly point-like particles. Some strings make up matter particles, others the virtual particles that carry the various forces. What determines which strings are at the heart of protons and electrons, and which make up the bosons and mesons and photons, is their frequency of vibration and the pattern of the vibration. What determines the particular set of vibrations that we see in the universe?

The theory holds that it is the shape and size of the dimensions of space that determines them. One of the radical features of string theory is that it turned out to be possible to construct a self-consistent theory, one which predicts gravity and the other forces and unifies them (Einstein's Holy Grail), but only if the number of spatial dimensions in our universe is 10. Of these, only three are large and those are the ones we sense and inhabit. The other seven are thought to be everywhere, but tiny, smaller than a ten-millionth of a billionth of a centimeter, or we would have detected them.

Where are they? They are thought to be curled up at every point of our 3-D space, in a direction we can't see. What shape are they? They are currently thought to belong to a huge family of complex curved spaces called Calabi-Yau spaces. To picture such a space, think of a very complicated pretzel, with a number of holes and with a doughy part that twists and turns and passes through itself many times as viewed in our space. Think of these fancy pretzels connected to our usual three-dimensional space at every point, like beads on an embroidered sheet.

Under this notion all matter and energy as we know it is actually built of vibrational resonances, cosmic musical tones, if you will, in these tiny, invisible, curled up spaces, distributed everywhere, constantly forming and reforming, shifting, vibrating, colliding and interfering with their properties determined by the shapes and sizes of these spaces. As a result what we see are electrons, protons, quarks and photons with particular masses, charges, and spins, that are the same all over our universe.

The proposal that there might be other universes that formed or are forming all the time flows from, again, asking the question, in the new light of string theory, why should our universe be the only one? There is no real reason to suppose it is unique, but no evidence nor any prospect of evidence to show that it isn't.

We can in turn ask, is the particular particle zoo we see in our universe the only possible one, is there only one permitted set of physical laws, and does that mean that only one version of Calabi-Yau space is permitted even if multitudes of other universes exist? String theory is too immature to answer that question yet. But some day it might. At the moment, it appears that other Calabi-Yau shapes are possible, and other universes might therefore have physical laws rather different from ours. As discussed above, it is not at all clear that if a traveler from one of those tried to visit our space it would survive. Perhaps even more important, a relatively small shift in physical laws might rule out any life in another universe. If, for instance, the ratio of strengths of gravity and electricity were slightly different in favor of electricity, presumably due to a differently shaped set of 7-D Calabi-Yau spaces, then stars might have to have more mass to radiate the same amount of energy. With a larger difference in these fundamental properties the stars could not shine, and such a universe would likely be lifeless.

The same sorts of considerations apply if we ask, what if the laws of physics vary from place to place inside our universe? The answers are those above. But we have clear observational evidence from the stars in our galaxy and from the distant galaxies that physics is the same everywhere, as Newton asserted back in the 1600s. Of course, it is possible that the laws of physics are everywhere different than we now believe, that string theory is wrong or that it is not the complete story. Such is always possible, but then we will have to await our own advances or the arrival of alien technology to find out.

ESCAPE TO THE TENTH DIMENSION

We seem to be left in a curious position. The most feasible way to walk through walls appears to be the detailed superfine subatomic control, which we are not likely to master any time soon, if ever, because of the sheer size of the herd of particles in our bodies and the wall we would have to control. Perhaps advanced aliens have learned to do it, if they have been around for longer than we have. The idea that aliens might be visiting us from the fourth dimension seems to be forbidden by the problem of how 4-D matter can exist in 3-D spaces. The alternative plan, of 3-D aliens jumping up-dimension to 4-D land, walking around the wall in a 4-D direction, and returning to our space seems to be ruled out, for a reason that pops out of string theory mentioned above: the fourth and all dimensions up to the tenth required by string theory, though they are everywhere "attached" to our space, are tiny and curled up. They are too small to hold a human, any alien of our size, or a spaceship. They are just large enough to hold single subatomic particles of our matter. In fact, that is what they do, in the form of vibrating strings and membranes that in our 3-D world look like particles. Even more curious, if this view held by string theory is correct, we are "really" ourselves 4-D (and actually 10-D) beings, all of us, all the time, though we walk and talk and love insensible to this state of affairs. In a sense this means that my suggestion that higher space beings might present a face in three dimensions which we could see is true: we are those beings! We are surrounded by extra dimensions, we are part of them, but we can't buy a ticket to visit them. We can't even find a published schedule for the trip.

7

UFOs and Abductions

THE NATURAL AND UNNATURAL HISTORY OF ALIEN INTERVENTIONS

The number of chronicled UFO reports from around the world had, by the arrival of our new millennium, reached the tens of thousands. In the United States alone, it has climbed into the thousands. A few computerized databases of UFO reports have been started in the last decade and some hold thousands of reports. Obviously, it would be impossible to analyze and report even a significant fraction of these in a book of modest length. Instead, I have chosen to try to summarize here the trends in both the reports and the reporting, and to discuss at some length the details and meaning of a few cases I find most significant.

The simplest statement you can make about UFO reports is that they lack hard evidence. Even most people inside the UFO community are willing to admit that, including some of the most prominent UFO researchers who have spoken at meetings on the subject held in different venues over the last fifteen years. See, for example, numerous instances in Randles' and Warrington's *Science and the UFOs*, the *1987 MUFON Symposium Proceedings*, Bryan's *Close Encounters of the Fourth Kind*, and Sturrock's *The UFO Enigma*. Having concluded that hard evidence is scarce, it is now our task to review the phenomena.

ALIENS IN ANCIENT HISTORY: EZEKIEL AND THE WHEEL

Over the years there have been allegations in literature that some of humanity's most ancient texts reveal that Earth has been repeatedly visited by aliens. Texts cited include the *Bhagavad Gita* from India, the *Polpul Vuh* of the ancient Maya, and the Book of Ezekiel from the Jewish Bible. In many ways, the latter account is the most suggestive, and it lends itself to a fairly detailed analysis. Recall that the prophet Ezekiel was living with a large portion of the people of Israel, who had been exiled to the banks of a river on the outskirts of Babylon. The text begins as follows:

> 1.1 In the thirtieth year, in the fourth month, on the fifth day of the month, as I was among the exiles by the River Chebar, the heavens were opened, and I saw visions of God
>
> 1.2 On the fifth day of the month [it was the fifth year of the exile of King Jehoiachim]
>
> 1.3 The word of the Lord came to Ezekiel the priest, the son of Buzi, in the land of the Chaldeans by the River Chebar; and the hand of the Lord was upon him there.
>
> 1.4 As I looked, behold, a stormy wind came out of the north, and a great cloud, with brightness round about it, and fire flashing forth continually, and in the midst of the fire, as it were gleaming bronze.
>
> 1.5 And from the midst of it came the likeness of four living creatures. And this was their appearance: They had the form of men,
>
> 1.6 But each had four faces, and each of them had four wings.
>
> 1.7 Their legs were straight, and the soles of their feet were round; and they sparkled like burnished bronze.
>
> 1.8 Under their wings on their four sides they had human hands. And the four had their faces and their wings thus:
>
> 1.9 Their wings touched one another; they went every one straight forward, without turning as they went.
>
> 1.10 As for the likeness of their faces, each had the face of a man

in front; the four had the face of a lion on the right side, the four had the face of a bull on the left side, and the four had the face of an eagle at the back . . .

1.13 In the midst of the living creatures there was something that looked like burning coals of fire, like torches moving to and fro among the living creatures; and the fire was bright, and out of the fire went forth lightning.

1.14 And the living creatures darted to and fro, like a flash of lightning

1.15 Now as I looked at the living creatures, I saw a wheel upon the earth beside the living creatures, one for each of the four of them.

1.16 As for the appearance of the wheels and their construction: their appearance was like the gleaming of a Tarsis stone [beryl]; and their construction was as though one wheel were within another.

1.17 When they went, they went in any of their four directions without turning as they went,

1.18 The four wheels had rims; and their rims were full of eyes round about.

1.19 And when the living creatures went, the wheels went beside them; and when the living creatures rose from the earth, the wheels rose . . .

1.22 Over the heads of the living creatures there was the likeness of the firmament, shining like rock crystal, spread out above their heads . . .

1.24 And when they went, I heard the sound of their wings like the sound of many waters, like the thunder of the Almighty, a sound of tumult like the sound of a host; when they stood still they let down their wings . . .

1.28 Like the appearance of the bow that is in the cloud on the day of rain, so was the appearance of the brightness round about. Such was the appearance of the likeness of the glory of the Lord. And when I saw it, I fell upon my face, and I heard the voice of one that spoke.

(Quoted from Blumrich, *The Spaceships of Ezekiel*, checked in Hertz, *The Pentateuch*.)

In this account it is not too difficult to get the impression that the apparition seen by Ezekiel might have been made of polished metal, that it arrived in a pall of smoke and mist with lightning bolts flashing in the cloud, with a brilliant flame at the center of the cloud, and with a sound like the roaring of a great cascade.

Blumrich, in *The Spaceships of Ezekiel*, elaborates the picture. He suggests that the prophet saw a craft shaped like a top, with a curved hemispherical upper surface, a narrowing, rounded-off, balancing nose, with a rim from which extend four retractable legs, each with cylindrical bottoms terminating in two narrow legs capped with disks. ("Their legs were straight, and the soles off their feet were round; and they sparkled like burnished bronze.") At the head of each cylinder he places a four-bladed rotating helicopter propeller, which only rotates as the craft hovers, flies at low speed or lands. Otherwise, they fold up (". . . each of them had four wings"). He interprets the lightning flashes as the flares of steering rockets employed as the spacecraft lands. The crystal firmament he visualizes as a transparent dome on the top of the craft. And the wheels are stowed at the bottoms of the helicopter units. They are deployed to allow the craft to roll around on the ground. The small wheels, which have their axes end to end around the perimeter, make up the large ones and allow the ship to travel one direction by rolling on the large perimeter made up of the small wheels' rims while they are held steady, and instantly to travel at right angles by energizing the small wheels. The various faces are a problem to interpret, but he suggests they are the best analogies Ezekiel could come up with to explain the appearance of the various equipment mounted at the tops of the legs, above the rotors. The "something that looked like burning coals of fire, like torches moving to and fro among the living creatures" he proposes is the power plant and exhaust, located at the tip of the underside. The overall design looks rather ungainly, but strip away the legs and the transparent dome and it looks rather like the shape of an upside-down Mercury or Gemini space capsule. Perhaps this is no accident because for many years during the heyday of the Apollo program Blumrich was a NASA engineer.

Blumrich, using later verses, puts together a whole scenario in which Ezekiel is visited repeatedly over a period of thirty years. The visitors look

This remarkable photo was shot from a moving train traveling through Oberwesel, Germany, in 1964. The disk rose into the air beside the train, with what appears to be a whirling energy vortex below it. Except for the missing four understruts, it is remarkably similar to Blumrich's vision of Ezekiel's wheel. Photo: copyright 2000 Fortean Picture Library

like and speak like men. They transport him in the ship from and to the desert, and up the mountains to the Temple, and in and out of the Temple.

The analysis and argument are extensive, but in the end, not convincing. There is too much traditional imagery related to living things and to angels, and it is simply easier to assume Ezekiel had visions rather than a visit from aliens. However, had there been a couple of eyewitnesses to corroborate Ezekiel's story, or had similar appearances been recounted biblically, we would be more inclined to take this account as a serious piece of reporting rather than a prophet's visions.

THE SHAPE OF THINGS THAT CAME; GEORGE ADAMSKI AT PALOMAR GARDENS; THE RANGE OF UFO SIGHTINGS

The variety of reported visible UFO shapes is fairly wide. In the period 1947 to the 1960s when seen as hard-edged objects, a large number appeared to be saucers, often with lower half different in shape from upper half, often with a dome, often with windows, frequently with a few or in some cases many lights, situated either under the rim or at the rim. These lights are sometimes white, sometimes red, green, and blue. Often the lights flash, and often the whole disk is seen to rotate around its axis. Sometimes (most frequently in abduction scenarios) the disk is seen to emit a light beam that is bright and unpleasantly actinic. Sometimes the beam shuts off some or all electrical equipment it illuminates; sometimes this effect occurs without the beam, simply from close proximity to the UFO.

A second major shape class from the same period is of cigar-shaped ships. These can be said to resemble dirigibles of the early 1900s. However, very few dirigibles have been flying since that time, and one cannot explain the numbers of sightings based on actual dirigibles in service during that period. Most observers claim that the cigar-shaped ships are larger than the saucers and have usually asserted that they are the saucers' motherships. These vessels sometimes have lights. In general, they and the saucers are described as having metallic surfaces. More recently the percentage of saucers and cigars has dropped and the number of wedge-shaped ships has increased. (See numerous ongoing reports in *UFO Magazine* published in England.)

A number of researchers have pointed out that the shapes of UFOs have tracked changes in military aircraft. In a rough way, this is true. The latest switch to wedges, at least, approximately coincides with the publication of images and models of the U.S. stealth fighter and B-2 stealth bomber. The bomber is a flying wing. Its shape is nearly identical to that seen as speculation on the covers of *Popular Mechanics* magazine in the 1940s and then in photos of Howard Hughes' actual craft of that period that flew a few times. It is a shape that has floated through the aircraft design community and popular literature since then. The F-117 stealth fighter has a distinctive faceted wedge shape that looks unlikely to fly, as it does not have a typical, streamlined jet aircraft profile. Yet it flies just fine. It is said to have the radar cross-section

This wedge-shaped craft appeared over Medjugorje, Yugoslavia, in 1991. It might be an experimental aircraft, or something completely unknown. Photo: copyright 2000 CUFOS

This formation of UFOs over Italy was shot in 1960. The craft have the appearance of the space shuttle seen head on, with the leading edges aerodynamically shaped to help provide lift while flying in an atmosphere. Photo: copyright 2000 Fortean Picture Library.

Shaped much like a child's spinning top, this UFO was captured on film over Rhode Island in 1967. It, too, is reasonably like Blumrich's scheme for Ezekiel's wheel. Photo copyright 2000 Fortean Picture Library

(reflectivity) of a pigeon, which, of course, was the point of the design. The image of the F-17 was shocking and quickly received wide publicity once it "came out" in public.

The earliest recorded cases of cigar-shaped craft in modern times go back to the late 1800s. These coincide more or less with the advent of both actual airships and fictional air adventure stories in the popular literature. The work of Jules Verne, among others, comes to mind.

The saucer shape as a type has been codified since Kenneth Arnold's 1947 seminal encounter, but the story is not quite so simple as often portrayed, yet perhaps instructive. All accounts agree that Arnold was puzzled by the fact that the craft he saw had no tails and no stabilizers. In one account, Arnold is reported to have made drawings for the Air Force that show the objects he saw as more or less boomerang-shaped (Randle and Schmitt, *UFO Crash at Roswell*, p. 13). According to this account, he told a news reporter that the craft flew through the air at incredible speed, holding to a chevron formation, while moving like saucers skipping across the surface of a lake. The news account published by the reporter used the term "flying saucer." In later interviews with the Air Force, Arnold appeared to have revised his story and described the craft as saucer-shaped. He made a crude sketch included in one Air Force document that looks in top view like a military badge with one rounded blunt end and an opposite end that is also blunt but with an obtuse peak, and it is very thin in cross section (Steiger, *Project Blue Book*, pp. 23-33). It didn't matter that it was not really a round disk in the usual sense. The term and image of the flying saucer stuck like glue.

It is slightly ironic that Arnold was, in effect, talked into this terminology. I think the reason the image traveled around the world is that it was a case of apparent authority putting its stamp on an already available image. Space ships had actually been a near fixture of the adventure-fiction world since 1898 when H. G. Wells published the invasion story, *The War of the Worlds*. (Wells based his work on Percival Lowell's telescopic observations of Mars and supposed Martian canals. This began the still-continuing interaction of science and science fiction.) In Wells's novel, the Martians arrive in cylinders and start mowing down human troops with their death rays.

In the 1920s, Hugo Gernsback started to edit and publish the first science-fiction pulp magazines and saucers, death rays, and little men with big eyes (some of them green) were all over the stories and the pulp covers. I am holding in my hand an issue of his *Amazing Stories* from January 1927, and on page 914 a lot of four-foot-high bug-men with big eyes are haranguing a pair of human captives. The images of saucers and little aliens were *available* in 1947. When the reporter interviewed Kenneth Arnold, the public perceived him as a trained pilot, articulate, careful in his assessments (for the most part), and a credible witness. His story went around the world.

Other aspects of UFO behavior, mostly related to abduction accounts post-1961, can be found in 1930s and 40s sci-fi pulps, and in late 1940s and 1950s science-fiction movies. I have a wonderful cover from a 1930 issue of *Amazing Stories*, showing the Empire State Building being lifted wholesale from Manhattan by a giant tractor beam. The death ray was seen in the 1930s Buck Rogers movie serials, knocking down cheesy rocketship models. I watched many of these on Saturday morning TV as a child. The classic movie *The Day the Earth Stood Still* featured an alien black box that could stop all earthly machinery. In the movie *This Island Earth*, a light plane is tractor-beamed up into a UFO and the occupants extracted and examined under restraint.

America has dominated world cinema since its beginning, so these images have been in wide circulation. It, therefore, seems quite fair to say that by the 1950s, at the latest, many standard components of both UFO sightings and abductions were a sort of common pop-cultural currency.

For some UFO sightings, there is additional evidence beyond eyewitness reports in the form of photographs and, more recently, videotape. My first brush with saucer photos was a set of photographs published by George Adamski in his 1953 *Flying Saucers Have Landed,* which my parents bought for me when it came out (I was eight years old at the time). I was quite convinced for a while that saucers had landed. Adamski's saucers looked like hubcaps on the bottom, in a general way, with three equidistant hemispheres protruding from the underside, and with a narrower cylindrical section punctured by rectangular windows on the top side; the cylinder was covered by a dome. This seems to be the sort of generic model for many photos and videos ever since. If you want to see more recent examples, look at the pictures reproduced in Jim Wilson's article in *Popular Mechanics* (July 1998), especially the McMinnville (1950), Trindade (1959), and Zanesville (1966) shots (pp. 62, 65, and 67). They are all variations on this single theme, remarkably similar in general shape (though different in details), even though the incidents were spread out over 15 years.

All share a common problem with one another, with Adamski's photos, and with most other UFO shots: the ships are seen against an open sky with no landmarks behind, and thus no way to judge scale. As some investigators have demonstrated, this is the simplest arrangement with which to fake a

Cover art from 1928. This spherical craft might serve as an interstellar generation ship. From the author's collection.

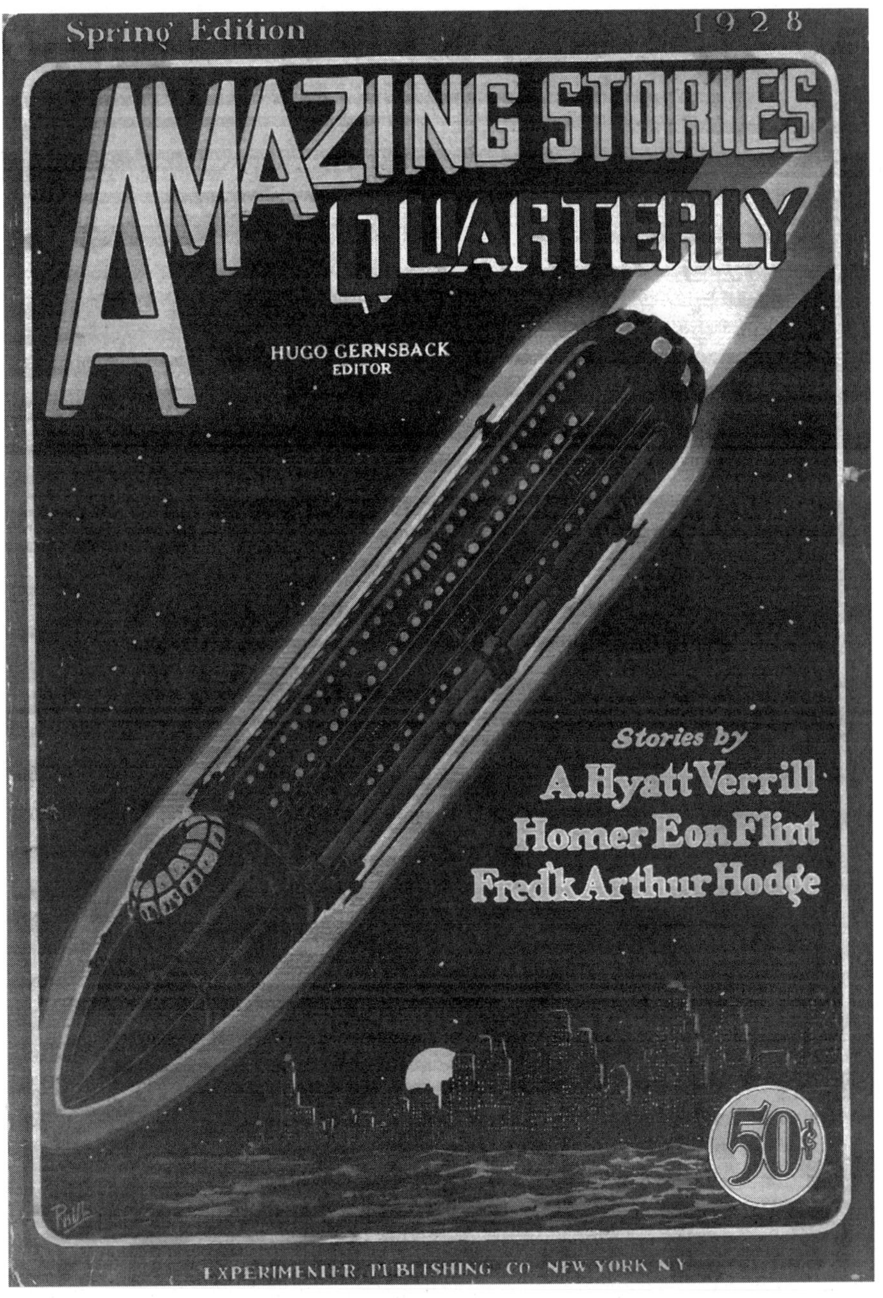

Another pulp cover from 1928, this one featuring a "cigar-shaped" craft. From the author's collection.

This 1935 pulp cover features exotic-looking spacecraft that three decades later may have inspired the world's most popular fictional starship. From the author's collection.

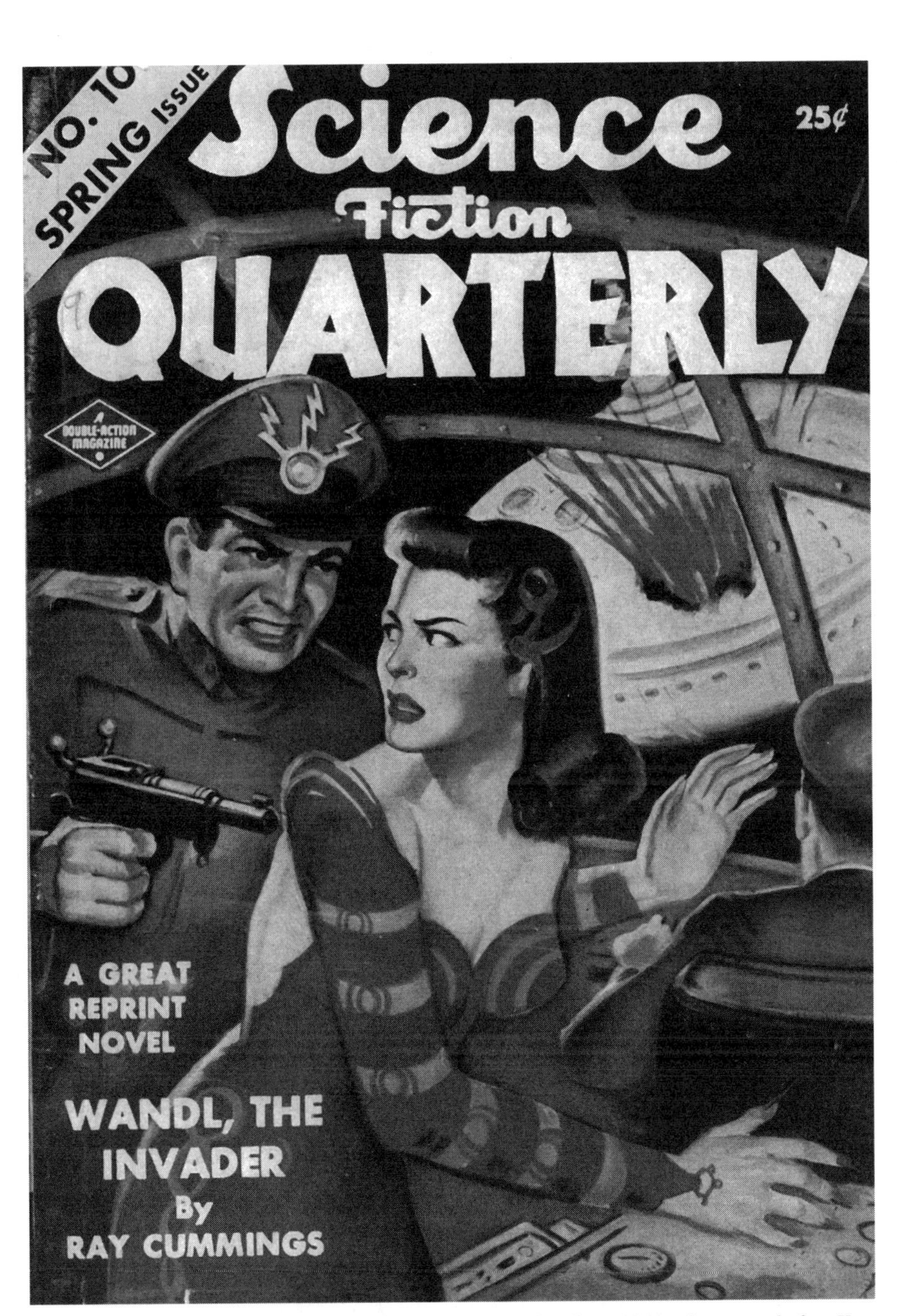

Flying saucer imagery appeared on the cover of this sc-fi pulp in 1943—five years before Kenneth Arnold's sighting. From the author's collection.

UFO shot; you can do it by tossing a small model into the air (or hanging it by a thread) and catching the model on film. The photos also share the problems of poor contrast and lack of detail. By the way, the second photo on page 67, of the Hillsdale "saucer" (also 1966) is an obvious case of lens flare due to the internal reflection of a bright setting Sun. The Sun's direct image, seen among the leaves of a tree, is so bright it is overexposed. Lens flare is given away by its typical shape, that of an ellipse with a spike on one side that flares into a sort of fan. The flare is equally far from and on the opposite side of the center of the picture from the bright source, and the spike and fan point away from the bright source. Lens flare might often pass unrecognized today because most decent lenses, even on cheap cameras, have high-tech multilayer antireflection coatings, specifically made to avoid reflections and flare. I recognized the flare in the Hillsdale photo because in my youth I had cameras with uncoated lenses and had taken pictures that clearly displayed lens flares. (You can spot the coating on your camera because it gives a faint blue cast to the front surface of the lens.) This clearly explains one of the "Six Unexplainable Encounters" of the article. It is surprising to me that the *Popular Mechanics* editors permitted the Hillsdale photograph to run. Though Adamski and his colleagues claimed for many years that his photos were genuine, recently published accounts have clearly shown the opposite.

A brief but amusing article on how to fake UFO photos by Cohen *et al.* was published in *The Skeptic.* The potential for fakery in photography has always been great. Even though evaluation tools for discovering chicanery in photographs have long existed and have become more sophisticated with the advent of computers, the scale has tipped far in the direction of computer-assisted visual fraud. I have committed such fraud myself, but just once. The year we returned to California my family wanted to send out a New Year's card showing us in a distinctive California landscape. The best current photo of us showed us standing on the lovely coastline of Big Sur. Unfortunately, I was wearing a white T-shirt with a noticeable rip at its V-shaped neck. I was not going to have our friends and relatives see me like that! So I scanned the image into a computer at school and, using standard software, changed my collar to a crew neck using pixel-by-pixel correction. Then I colorized it to a nice faded green. The final result is almost perfect even though it was my first attempt and I had no formal training in the process. When it comes to UFO photos, *caveat emptor*!

The same must be said of videotapes of metallic saucers and other definite shapes. An excellent example is the mass daytime sighting that took place in downtown Mexico City in the 1990s and was viewed by thousands. A craft was seen to hover and move slowly above the buildings. I recently saw a video of the event and the ship looks something like a blimp with hard edges and excrescences. It appears to glide slowly behind several skyscrapers. The UFO is a low contrast object in a low contrast image. It drifts behind buildings so it is possible to estimate a minimum set of dimensions based on it being just behind the buildings. The result is that the craft was at least 300 feet long and 75 feet in diameter. If it was much farther away than just behind the buildings, it had to have been much bigger. It is impossible to verify the size and impossible to prove that the image is either authentic or a fake.

The largest number of UFO sightings report lights in the sky that have no obvious surface or shape. They are generally self-luminous, and often come in groups that fly in formation, as in Arnold's original sighting. Not surprisingly, the UFOs seen with shapes are generally daylight events and the lights are mostly seen at night. These lights, along with radar sightings, are the UFOs most likely to abruptly change directions and speed off at impossible accelerations. They are almost always silent. The numerous photos of such lights tell almost nothing. They are inevitably white blobs of indeterminate shape on a black background.

A recent example shows more than that but is just as problematic. In his article in *UFO Magazine* (June 2000, pp. 22ff.), Bill Hamilton reproduces frames from a video supposedly taken of the Phoenix Lights of 1997. In the first image on page 23, a set of seven bright spots forms a dashed arc across a dark sky above the lights of a city and the silhouette of a foreground hill. These lights were reportedly seen by many people on the ground. You can tell from the picture that each spot is just a few pixels across, since the pixels are obvious. A shot on page 25 shows supposedly the same set of lights in the same situation, but, in this case, they are in front of a distant set of mountains not visible in the first image. The article and captions are a bit unclear, but it appears that the second frame was a superimposition of a night frame of the lights and a dusk frame of the countryside made in an attempt to determine the distance of the lights. Only this time there are only five luminous spots.

The article attempts to decide what these lights were. One explanation is that it was a set of aircraft flares dropped from high altitude more than fifteen miles away. A set of much clearer images made from photos instead of video is reproduced on page 27. These show the descent of a group of four bright objects above a distant mountain range. Their caption says, "Flares?" and the objects *do* look a lot like flares. They have a bright center, diffuse edge, and what seem like obvious smoke trails distorted by winds aloft. The author tries to argue that the lights did not look like flares because they were sharp-edged and golden rather than white in color. But at a great distance, the edge would look sharp, the smoke trail might be too faint to be observed and the color would tend to the yellow due to atmospheric light scattering. It is a typical, if unresolved, case.

Radar sightings of UFOs sometimes alone and sometimes coupled with visual sightings have been reported since 1948. These sightings have been made mostly by military personnel who, it has been pointed out in the literature, have no stake in fakery and could suffer for it. They are also technically trained, and, in many cases, experienced observers. Radar operators are trained to look for intruders, and when an unscheduled blip appears on the screen, changing directions abruptly at many g's and traveling at speeds above mach 3 (back in the 1950s, above mach 1.5), they take notice. In some cases, an equipment malfunction has been discovered, which explains the blips. In other cases, it can be shown that atmospheric conditions are conducive to peculiar reflections and scattering of radar pulses. The most usual and symptom-producing type of condition is called electromagnetic ducting. This occurs when the air is hotter above than below (which is called an atmospheric inversion), and the sharp change in density at the boundary between the layers acts like a radio mirror. This, in effect, forms a slab of air, trapped between the ground and the hot layer, in which the radar pulses can reflect back and forth. If there is a distant object, such as a mountain range, or even a sharp weather front with an abrupt discontinuity, the pulses can bounce back strongly from it. They can come back singly or as multiple images. The multiple images might appear to fly formation, rising and falling abruptly, or scatter at random, depending on the pattern of winds and boundary layer motions in the duct. This phenomenon has been amply documented, but it has not been clearly demonstrated to be the explanation for any particular UFO incident.

The cases where there is a visual sighting simultaneous with a radar blip are not so easy to explain, but might be due to a similar thing. If the atmosphere bends light (or radar) so much that it curves down toward a planet's surface faster than the surface curves, the atmosphere is said to be superrefractive, and light or radar can bounce back to any place from any other place on the planet. This is the natural state of affairs on the surface of Venus. On Earth it is more rare, but ducting can make it happen. Certain optical mirages are created this way. The Arctic illusion called the "fata morgana" is created by ducting, making distant ice surfaces appear to the viewer to float in the air as white columns and spires (see Sturrock, *The UFO Enigma*, p. 141). The conditions for ducting are improved by the presence of water vapor, which is also the condition for clouds. The presence of clouds (which scatter and absorb light) will often mask the optical illusion but leave the radar echo untouched.

There is one group of reports that is particularly striking due to the phenomenon itself and the putative expertise of the observers, which may also have an interesting natural explanation. These are the reports by experienced pilots of being tracked by luminous spots that station-keep with their aircraft. Since the 1940s, these aircraft have ranged from prop planes to jet fighters. Both day and night incidents have been reported. Night versions of these lights, called "foo-fighters" at the time, were seen by Allied air force pilots during World War II bombing runs.

One experienced meteorologist reported that a bright white light kept perfect station between his plane and the ground for several minutes. The sky seemed perfectly transparent above and below his plane. But by checking the angle between his eye, the spot, and the Sun, he was able to show this was due to a strange reflection called a "subsun," caused by flat transparent ice crystals, falling so that their surfaces remained horizontal all the way down. They acted as transparent reflectors. It may be that the UFO night cases and other day cases could be explained by flat ice crystals falling with other orientations, and by reflections of other sources than the Sun such as the Moon, or airplane lights. (See Sturrock, *The UFO Enigma*, p. 140.)

In spite of the collected statistics and the attempts to systematize data collection, we are not much closer to resolving or understanding these reports than we were fifty years ago. Governmental efforts have been made in this area in Norway, Sweden, and France. In Sweden, the Hessdalen Project has attempted to develop a set of at least two stations with automatic TV cameras and spectroscopes, to allow automatic recording and triangulation of unusual optical events, principally the bright lights of unknown origin that have sometimes been tracked on radar at 18,000 miles per hour as they traveled down the Hessdalen Valley near Trondheim (see Sturrock, *The UFO Enigma,* pp. 78-80). Project URD and Project UFO-Sweden have an elaborate multi-page form for data collection and an organization originally government supported, which included an educational component in the schools. The data presented show that the number of UFO incidents went from zero in 1947-56 to two in 1957 to a peak of 119 in 1979, down to zero again in 1986. (This data is reported in the *1987 MUFON Symposium Proceedings*, pp. 156-164.) Unfortunately, even with all the elaboration, the data set is not really useful except to tabulate totals, dates, locations, and trends in incident descriptions.

The largest effort was the French GEPAN/SEPRA Project set up in 1977 by CNES, the French national space agency. The great advantage it has over the others is that it was able to command the making of incident reports by the gendarmerie, meteorological offices and national police and have them sent back to GEPAN. This has created a large data set collected by neutral and somewhat trained reporters. This effort was coupled with a research program on exotic propulsion technologies, basic research on unusual atmospheric phenomena, development of image analysis techniques, and special attention to aeronautical incidents, especially ones that are both visual and radar (see Sturrock, *The UFO Enigma*, pp. 131-135). Of about 3,000 reports collected under this system, about 100 were judged to be worth careful investigation. Of these only a few cases are still unsolved. The effort has been ramped down, though there is still interest, and a plan exists to build automatic tracking stations linked to computers, but as yet there is no funding.

DECIPHERING THE MEANING OF THE LIGHTS IN THE SKY

I would like to suggest some ideas that might permit progress in evaluating the sort of optical and radar reports discussed above, which almost uniformly suffer from lack of physical evidence. Witnesses should collect data at the time of an incident or as soon as possible afterwards. They should use, if possible, a standard outline or set of questions to answer so that data are as comparable as possible. In fact, these steps have been taken by a number of researchers in the references cited.

A first step for new researchers would be to study the existing literature, some of which is referenced at the end of this book (and in the bibliographies of some of those references). Familiarity with the range of the frequent phenomena should be useful. Then they should try to learn about atmospheric and other natural phenomena. Two excellent references are: Greenler's *Rainbows, Halos and Glories,* and Minnaert's *Light and Color in the Open Air.*

Careful thought should be given to what instrumental evidence would be most useful and how best to record it. Look at the suggestions in Sturrock's *The UFO Enigma.*

Portability would be a plus unless you have a site you think is favored by lots of sightings. Having two cameras or camcorders would be useful for triangulation. To make them portable, you could mount them on the widest arm you can manage and arrange to have them expose simultaneously. If possible, date- and time-stamp the images automatically. Be sure to pan the cameras side to side to establish the angular rate of motion of moving objects and the distances to fixed ones whose sizes and distances you measure (by map or pacing off) before or after the event. Be sure to make images of objects of known size before you shoot a UFO during the sighting, if possible, and afterward. An audio recorder would also be useful. It might be possible to use a hand-held radar gun to measure speeds and perhaps ranges. A cheap laser pointer shined at the object could show whether the thing is a material object, by the presence or absence of the beam's reflection. With a beam chopper, photo detector and timing circuit, or maybe just detection of the reflection by the video cam, another determination of the range of a UFO might be made.

There is a continuing mystery about the nature of the UFO lights that lack form or surface. Several possible natural explanations might be ruled out by fairly simple observations. It has been suggested they might be due to dielectric effects caused by shifts along geological faults, resulting in temporary electric field effects above Earth's surface. Some people have reported seeing lights near the ground just before earthquakes. These may well be real, physical external phenomena—perhaps a form of ball lightning. It is also documented that some animals become disturbed just before earthquakes. It might also be that these dielectric effects are stimulated by external fields but are internally experienced as the result of direct influence on the nervous system. In that case, people should report seeing lights but video and camera shots should fail to record them. That is rather negative evidence, though worth going after. To further check that this is the case, one could take advantage of the fact that a closed conducting surface will shield the space inside from external electric fields. This fact is made use of in essentially all cabling that carries electrical signals. This shielding does not require a continuous sheet of conductor; it can be provided by a metal screen, as long as the screen mesh is smaller than the wavelength. Standard insect screen would work against all radio and even microwaves. Such screen enclosures, which are called Faraday cages (after their inventor, the famous 19th century physicist, Michael Faraday), are used to shield "secure" rooms in embassies and spy agencies around the world. I suggest you make a screen (that you can see through) or an aluminized plastic film with eye holes cut out, large enough to enclose your head at least, perhaps your whole body, and "e" field. You would put it on as soon as you see the lights. If they then disappear, and reappear when you remove the cage, then direct influence on your nervous system is a good possibility. It would obviously help to do this with a companion for verification and assistance.

There is another potential benefit to wandering around with portable Faraday cages: it might prevent shutdown of electrically driven equipment or damage to you. The prototypical on-the-road close encounter with a UFO involves the failure of the car engine, lights, and radio (including the radio transmitter, if it is a truck or police car). It is possible this could be explained by some sort of strong magnetic field. If this is the case, then additional Faraday cages could disclose more information. A cage surrounding the engine block might shield it and keep it running, along with the lights. A cage around the radio would not help, because the cage would itself shield both incoming and outgoing transmissions. A cage around a geiger counter should keep it working; it receives subatomic particles, not radio. (But, curiously, there are reports in which geiger counters are not affected while generators and cars are. This inconsistency makes such accounts doubtful, or at least much harder to explain through known physics.)

There have been cases in which the observers have come away from a close encounter with rashes, superficial burns, and sometimes blistering and other skin problems. Some have been evaluated for nuclear radiation exposure, so far with negative results. A few researchers have suggested as an alternative that these victims were exposed to microwaves. The reason any spaceship would radiate high intensity microwaves is obscure. But if it is the case then a Faraday cage should also shield the observer from such effects. (Cages are incorporated in all microwave ovens to shield users from their radiation.)

OPTIMUM SPACESHIP DESIGN?

In thinking about the UFO zoo of reported shapes, one can ask, do these make sense in terms of known design principles? Take the case of ships meant only for travel in the vacuum. (In the UFO world, the motherships.) In principle this is the easy case. There are no major external design requirements. The ship is not intended to land on a heavy planet or travel through an atmosphere. Thus it can have any shape, grow like Topsy, be built of relatively fragile materials, and be as large as other constraints such as money and time permit. This fact has been admirably envisioned in many sci-fi movies and includes flying cities, towers, tinker-toy constructions, and in real life, the ungainly Apollo Lunar Landing Module, which looks like an insect. The design logic would be dominated by considerations of ease of construction, internal travel, views to the outside, storage access, cargo manipulation, repair access, egress and ingress, waste heat dumping, etc. From this point of view, the cigar-shaped motherships so often reported are over-streamlined and not particularly plausible. Of course, a mothership could have any shape and still land on a planet if it had force shields and tractor beams, in other words, if it controlled space warps, as discussed elsewhere.

For "scout" vessels that routinely descend to planetary surfaces without such force fields, the designs should be dominated by the requirements to fly smoothly with minimum drag, maximum lift, and good natural stability and maneuverability. Thus human airplane designs are what we would look to for examples. For slow flight, cigar-shaped craft would be satisfactory and of course dirigibles are the main example from Earth. Their design represents the congruence of the flight requirements for stability, modest maneuverability, and the maximization of gas storage capacity and rigidity along with minimization of structure.

For high speed designs look to jet fighters, including the new Stealth fighters that embody a wedge shape with thin edges, which is now in UFO vogue. The pointy shape gives the needed front/back asymmetry that enhances lift and lowers drag. The question is why would anyone design a saucer-shaped high-speed craft? It is not inherently a particularly low-drag or high-lift shape. It has no front and back and would be hard to recover from various kinds of tumbling and stalls. It is not inherently stable. The only aerodynamic way to enhance lift and improve stability for a saucer is to provide it with rapid spin. The Frisbee® flies because of its spin. (Try throwing one without spin, and see what happens!) It is true that many UFO reports tell of spinning saucers, but the rates of spin reported are generally too low to provide aeronautical benefits.

And then there is the question, how do you see out of a ship when it is spinning rapidly? Almost all UFO saucers with windows, and there are many, including those photographed, which have the windows on the rim. In fact the aerodynamically required spin rate would blur out the window and other detail seen in many photographs, certainly those taken pre-1970 when really fast film was not yet available. This last fact, along with the illogic of the design, makes one suspicious of all the saucer reports. Of course, with force field shielding, designers can indulge themselves. Humans have always considered the circle a "perfect" shape. Perhaps saucers have religious or cultural significance for aliens.

And then there is the question of scale. Saucers generally have been reported in two size regimes: gigantic and tiny. A small minority, including Kenneth Arnold's original report, are of dimensions from 500 to 2,000 feet in diameter. These are sizes clearly impractical for—and with severe consequences to—on-planet travel, as discussed elsewhere. The vast majority of reports are of tiny ships, 15-30 feet in diameter. Even with crew who are four feet tall, if they are alive and well, this just seems too small. Where is the space for on-board consumables? I suppose with a matter transmitter or a replicator that draws matter from the surrounding atmosphere and planetary surface (or from the mothership), the actual food, water, etc.—or the raw atoms for their production—would not have to be carried on board. Otherwise, where is the beef?

Even more extreme is the fuel or energy requirement. Landing and taking off from a planet is energy-expensive. Even if aliens possess artificial anti-gravity drives, where do they store the energy to run them? If it is antimatter they are using (about the most efficient alternative), where is the space for the strong electric or magnetic fields and attendant structures that should be needed to contain and manipulate it?

There is another piece of behavior (so far unreported) that UFOs should undergo if they make high speed entries and exits from the Earth's atmosphere. This is the creation of temporary ionization trails from the craft's collision with air molecules. These should show up on radar as narrow reflecting wakes, just as observed for incoming meteors. In the right circumstances, admittedly somewhat rare, these wakes should be directly visible to us down here. The trails will be visible if the craft enters the upper atmosphere in a clear dark sky, near sunrise or sunset. I saw the Space Shuttle enter just before dawn one morning from the steps of my house in San Francisco, and I will never forget it. It came over the northwest horizon heading south-southeast, for its destination in southern California. The trail was as wide as about one-third the diameter of the full Moon. The color was a pale violet, like a faint welder's arc. The passage was completely silent. The tail end of the wake did not begin to disappear until the Shuttle had passed the opposite horizon, and then it just slowly faded without breaking up from upper atmospheric winds. It was really beautiful.

At the other end of things, the motherships should be really big and in orbit around Earth to support scout ship operations. Why don't we see them? We have powerful radars constantly looking for incoming targets. They can detect warheads as small as a foot across at 150 miles up. So they could likely detect an object 1,000 feet long at a distance of 100,000 miles. We also have got a fragmentary space tracking network looking for near-Earth-passing asteroids, and this can see targets of that size millions of miles out.

True, "stealth" technology could reduce the radar cross-sections of large ships. But they are still solid bodies and would visibly occult stars on a regular basis as they orbit us. These occultations should surely have been observed by now, given the constant observations of astronomers, both professional and amateur, and more recently the high-speed computer-controlled telescopic sky surveys now underway to detect supernovae and to look for gravitational lensing events in our galactic halo.

There are two logical explanations for why motherships have not been found by any of these detection methods:

1) These alien craft not only have radar invisibility but the ability to turn transparent;
2) Despite the reported sightings, they do not have physical reality.

CRASH LANDINGS AND ALIEN ABDUCTIONS

I have thought a great deal about why the Roswell incident has had and continues to have such resonance among the American public. I believe the first reason is its proximity in time to Arnold's seminal experience and the attendant publicity around that. The image of saucers was alive and sensational (even if the term was coined by a reporter from Arnold's description of the crafts' behavior, and the vivid visual associations of the public really flowed from the stream of science fiction).

The appearance and the likely actuality of government cover-up also has a lot to do with the long legs of this story. At the time of Roswell, the American public had just learned that its government had completely concealed for five years the most terrible and largest secret of the century: the development of the atomic bomb (the Manhattan Project). The climate was thus prepared for the public to believe in the possibility of another cover-up. More recently the unanswered questions surrounding the Kennedy assassination, the cover-up in the aftermath of the Watergate scandal and the Iran-Contra affair, and the questions surrounding Waco have revived the public's sometimes-justified paranoia and willingness to believe in cover-ups at the highest (or lower) levels of government. Third, there is a group of writers and talk-show hosts who have made it their business to amplify this fear at every opportunity. That is not a difficult task, considering that the army's initial press release on the Roswell incident stated in black and white that they had recovered wreckage from a crashed UFO.

It is the job of military (and other) intelligence to protect resources and assets. Without knowledge of the historical intelligence context, it is hard to understand the logic of censorship at the time of an incident such as Roswell. But a smart intelligence officer might learn a lot from a small and seemingly innocent slip, and thus we should have some sympathy for the intelligence devils of the time as they struggled to come up with cover stories for Roswell.

The government has recently stated that a cover-up at Roswell was necessary to protect Project Mogul—an experimental, high-altitude balloon designed to overfly and observe the Soviet Union's atomic-development program—and perhaps other, as yet unrevealed, projects from prying enemy minds. If this is so, the government has only itself to blame for the continuing controversy, because it has never issued an explanation of the thinking that went into using "crashed UFO wreckage" as the original cover story. This choice doesn't qualify as the product of rational military thinking because, given the climate of the times, such an admission was guaranteed to create a firestorm of fear, curiosity, and publicity. It would have been easier and wiser to simply delay a public announcement until the cover story of a weather balloon had been concocted. The lack of a rational explanation for the use of UFOs as a cover-up has helped to keep the controversy alive.

In the end, it is striking to me that no hard evidence of the actual artifacts and bodies reported to have been recovered at Roswell has ever turned up in any form. In the more than fifty years since the event, surely something or someone would have shown up who could have authenticated this story. Israel ran an atomic weapons program in secret for about thirty years. Despite the fact that they have enormous reasons for keeping it secret and despite having what many consider the best intelligence service in the world, eventually an idealistic young man succeeded in photographing the weapon assembly line and smuggled the pictures out for publication in England. He was caught and imprisoned, and Israel to this day does not acknowledge its program's existence. The breach, however, occurred.

A few years ago a video appeared on television and then on the market called *Alien Autopsy*, purporting to include film made in 1947 at the autopsy of the alien victims of the Roswell crash. After the film is over, the tape continues with interviews with "experts" who give their opinion of the authenticity of the film. Among them Stan Winston, the special effects wizard, contends that his crew could not duplicate what is seen on the film. This was disappointing because recent stories (see *UFO Magazine*, March/April 1999) show who was responsible for the fake and how it was done. You can read about the technique there, but the obvious problems I saw were in the internal details of the body as revealed by the pictures. There was no hint of a circulatory system, there was no internal body cavity, there were too few internal organs and they looked unreal, and the brain dissection revealed no stem to connect to the rest of the body. I use this film in my classes about the search for extraterrestrial intelligence (SETI), and survey my art students, many of whom are film and animation majors, as to whether they think the film is real. Their opinion each time is that not only is it fake, but that they could do better. *Alien Autopsy* is entertaining, but it is not real. Of course, that does nothing to settle the many open questions surrounding Roswell.

This brings us to the phenomenon of alien abduction, an increasingly loud and alarming part of the UFO scene. There are two cases I think are most important: the prototypical story of the 1961 abduction of Betty and Barney Hill, and the 1989 abduction of Linda Cortile.

Accounts of the Hills' abduction may be found in a number of sources (see the Recommended Readings for discussions that include several varieties of material). In brief, the Hills were driving home late one night in New Hampshire when they stopped to investigate what at first seemed to be a luminous object that had landed in front of their car. Their next conscious memory was of driving down a different section of road, miles from where they had stopped. Looking at their watches they realized it was two hours later, and they could not account for the missing time. Eventually their unease over the event, and especially the missing time, grew and they sought therapeutic help. They were separately hypnotized and regressed to the date in question. The doctor who saw them instructed them not to discuss their memories with each other. Each then remembered that Barney had seen a craft with windows, and through his binoculars had spotted beings inside. After it landed and the Hills got out of their car, they became scared and tried to flee. Barney was very fearful and this came through in his agitation during hypnosis as he related the events. The Hills were then led against their will into the spacecraft. There they were placed on tables and given medical exams. A needle was inserted into Betty's abdomen for what the alien in charge said telepathically was a pregnancy test. A sperm sample was taken from Barney, though this fact was withheld from the published accounts for many years because Barney found it too humiliating to reveal in public. Skin and hair samples were taken. Then they were taken back out to their car and their memories erased by a "beep."

This incident has received wide circulation. Even bigger have been the accounts of Travis Walton, seen on TV, and the best-selling books *Communion* and *Transformation*, in which Whitley Strieber recounts his alleged multiple abductions. By the 1990s, abduction accounts had also repeatedly surfaced on TV. But the Hills' story has virtually all of the elements seen in most other abduction stories: The abduction against their will, the sampling of body tissues, the invasive and often painful procedures made against their will, the focus on reproductive samples, and the telepathic communications with the aliens, as well as the recovery of lost time memories by hypnosis.

The lack of hard evidence is the main continuing problem. The very best case of a supposed "alien implant" during abduction was examined by Prof. David Pritchard of MIT (reported in Bryon's *Close Encounters of the Fourth Kind*, an account of the 1992 MIT conference on alien abduction). It supposedly came from the penis of Richard Price, who claims to have memories of its implantation there by aliens. The examination revealed nothing remarkable. The "implant" seems to be a small nodule of collagen and fibers that might have grown naturally inside Price. There was no way to prove that it came from aliens.

Attempts have been made to study the accounts, the backgrounds of the abductees, and the backgrounds and methods of the debriefers, for insights into the experience—all without consistent conclusions. Analysis of the accounts, especially those including the forced and painful tissue sampling or extraction of eggs and sperm, has shown they have close parallels to child abuse. It is not crazy to suggest that the accounts of aliens really mask true and older events of child abuse, mostly at the hands of parents. But the frequency of the accounts of aliens is rather high in comparison to documented cases of abuse, and in only a minority of cases of alien abduction has child abuse been documented. Also, not all alien abductions are negative experiences. In some cases, abductees report post-experience exaltation.

One of the presenters at the MIT Conference in 1992 delivered a number of statistics based on a sample of 32 abductees. There were three times as many women as men; the median age was 38 (probably meaning at the time of the survey); 94 percent were Caucasian, 58 percent were married; the average education was two years of college. On the average they had 1.9 children and 3.1 siblings. Most said they had been abducted between 1970 and 1979, meaning most had been young adults at the time. They were tested on the Index of Childhood Memory and Imagination, and the group average agreed with the general population normal. They scored about 25 on the hypnotic suggestion scale compared to about 20 for the general population. About 20 percent reported having occurrences of vivid images or sounds as they fell asleep or woke up. Their results on the Minnesota Multiphasic Personality Inventory (MMPI) were close to normal. The investigator also stated that detailed analysis showed two sub-groups in the sample. Group II scored higher on the MMPI and Keane-Post-Traumatic Stress Disorder Sub-scale, and it revealed greater incidence of sexual abuse when they were children.

What are we to make of these data? No definite conclusions follow. However, something is almost certainly going on at an internal psychological level. But what level? Bud Hopkins, an artist and UFO researcher, entered this field two decades ago and has collected many abduction accounts, and for some time has been running abductee group sessions as a combination of debriefing and a sort of therapy. He has published a number of well-known books detailing these abduction accounts.

Through Hopkins, Professor John Mack, a Pulitzer-prize-winning psychologist at Harvard, has joined the field and published *Abduction*, which relates in detail a few of the case studies he has made by counseling abductees. He entered the field with enormous authority and has switched over most of his practice to abduction work. He insists that the accounts must be real to the "experiencers," in part because the accounts he heard agreed in detail with one another and were from people who did not know one another and thus were independently derived. He has been careful to separate himself from those who absolutely believe in the physical reality of alien abduction. But in both his book and at the MIT Conference he does so rather coyly. He suggests over and over that there *are* visitors here, not necessarily from outer space, but perhaps from another reality or another dimension.

However, I have to question the asserted independence of his abductees' accounts. First, there is the fact, already mentioned, that by 1990 published accounts were widespread in various media, so even if the abductees he saw did not know one another, they were connected to the protoaccounts consciously or unconsciously. Second, I remember reading a short article by a journalist who was interested in Mack's work and in abduction, and decided to try to join his group by creating and describing to him a partial memory of an abduction. She wrote and called him and he graciously offered to have a preliminary meeting to discuss her case. A short time later but in advance of the meeting she received a packet in the mail from him that contained clippings and excerpts of some of his cases that described abduction scenarios in detail.

When abductees report being paralyzed by aliens, is there any other explanation? Many of us have experienced the hypnogogic state, at the edge of sleep, when we may feel as if we suddenly "fall" from floating above the mattress. This is a sort of muscle relaxation spasm, quite common, and I have felt it myself. A more extreme example that reproduces more of the abduction experience seems to have been around for a long time, thought it is not widley known. It is called "the Old Hag" and was thought to be a traditional folklore belief in many countries including those as diverse as America, Denmark, and Poland. The experience is described, fairly uniformly, as follows: There are footsteps in the bedroom, a shadowy presence can be seen, a weight climbs on the bed, a shadow crawls onto the person's chest, paralyzing him and squeezing the breath out of him. Sometimes the visit is accompanied by a dusty odor. In one case, the shadow spoke in an unintelligible tongue that nonetheless stimulated images in the sleeper's mind. The weight climbs off and the presence vanishes. In some cases, the sleeper becomes sexually aroused. The room is empty when the sleeper gets up to investigate. The amount of time this takes can vary from minutes to an hour.

The explanation offered by sleep experts is that this experience occurs when a person awakens (at least partly) from a state of sleep paralysis, and hyperventilates. This can reduce oxygen to the brain, which can induce hyperacuteness of hearing, so that faint noises sound loud. This in turn can induce hallucinations, including ones that encompass the sexual pleasure centers of the brain in both men and women.

This has all the earmarks of the visit of a succubus, that seductive creature of legend. It also has so much in common with alien abductions that one is tempted to say that many abductees (at least the ones taken from in bed) have been through a version of this real physiological state.

This would be more or less the last word were it not for the 1989 case, reported by Bud Hopkins, of the abduction of Linda Cortile. Cortile described being awakened during the night by short gray aliens in her bedroom. The room was illuminated by a bright shaft of light that lifted her, paralyzed, and wafted her along with the aliens through her closed apartment window, which is on the twelfth floor of a fifteen-story building in Manhattan, and into a waiting ship hovering just outside.

What makes this account special is that fourteen months after the abduction, two private security guards contacted Hopkins and stated that they had been on Manhattan's FDR Drive that night transporting an important unnamed personage when their car's engine failed. After stopping they noticed a glow ahead of them and looked up and saw an oval disk hovering outside an apartment building. Suddenly it turned on a blue-white light. A minute later three short aliens with large heads floated out through a closed twelfth-story window along with a woman in a nightgown. (They happened to have binoculars and so saw a fair amount of detail.) The figures disappeared into the ship, which promptly glowed orange and dove into the East River. Five months later still, an elderly women wrote Hopkins that she had been driving into Manhattan on the Brooklyn Bridge on the same night at the same time when the bridge lights went out and her car's engine and those of several others failed. She too saw the ship and the beam of light and the figures drawn out of the apartment building.

The woman contacted Hopkins after having seen a television special about him and the abductees with whom he worked. It is not clear how the security guards got to him. This case is so striking because of the external witnesses who had no personal relationship with the abductee. But were their stories independent? It is impossible to tell. Given the passage of time, the details of the scenario may well have been publicly available. What would their motives be for falsifying their stories? I don't know. However, some aspects of these stories could be checked out, but apparently have not been. One could check up on the guards' presence on the job that night, on whether the bridge lights failed, and whether the building facade involved can be seen from the places alleged by the witnesses. There is work to be done here.

What can be done to settle the question of alien abduction? I would suggest the following. For those who believe they are the victims of repeated and continuing abductions, hang a video cam in your bedroom, hooked to multi-hour videotapes. In addition, go on the web and create your own abduction website with a live webcam hung next to your video cam. Make sure a working calendar-clock is visible in their fields of view. Every night turn on the systems when you go to bed. Invite the public to watch your site. Who knows, perhaps nothing will happen, perhaps a clear case of sleep paralysis will manifest itself, and just perhaps, short gray aliens will appear on a beam of light and take you away.

OH, MIGHTY ARE THE WORKS OF MAN (AND ALIENS?)

At the outlying edges of the UFO phenomenon there are a number of so-called mysteries often linked with them. And there's a cottage industry that does the linking. I will briefly summarize the potpourri here.

The first and most heinous is the case of animal mutilations. The literature on this is rather impoverished, but it includes reports of sheep and cattle found dead by the roadside with organs and parts of their anatomy removed in ways that supposedly can't be accomplished by humans. What is going on? I suggest that humans are doing this. In cases where the animals are the property of someone and slaughtered without permission, there ought to be a criminal investigation. There could also be an investigation for disturbing the peace. Animal mutilations (though usually cats and dogs) are conventionally thought to be perpetrated by people who often turn out to be a certain type of serial murderer. It seems unlikely here because the multiple mutilations and the size of the animals suggests the work of a group of adults rather than of a single child, and they have not been associated with any human murders.

Of a more benign cast are the cases of crop circles. In general these are generic patterns impressed into fields of grains. In the last two decades these have become celebrated, especially in England. There are those who claim that they must be the work of aliens. Their evidence for this is, I believe, rather limited. They claim that humans could not have made them because the grain stalks are folded down, lie in complete order, and are neither crushed nor broken. And the huge and complex designs are executed overnight, and because the composition of the soil in these patterns is somehow unusual (without real specification). Thus only aliens can have produced them. Lucy Pringle, in her beautiful picture book *Crop Circles*, maintains just that. She labels as pretenders the two English farmers who stepped forward some years ago to claim credit for at least some of these. She concedes they might have produced some of the circles, and that other humans might have produced some of the others, but that the human-produced examples should be labeled frauds, and that such cases can be distinguished by the state of the grain from the true cases made by aliens.

Why go to the trouble of invoking aliens? Even if the technique used to produce each circle is still not known, it seems clear to me that these are human-made and many are quite beautiful, and eventually the artists ought to take credit. The designs for the most part come out of human traditional images, simple geometry and published examples of higher mathematics such as fractals. There are many interesting ways to create land-designs, as shown both in Pringle's book and one by Stan Herd called *Crop Art*, which showcases his work using a range of techniques. He spells these out to some extent, and they include guided planting, guided harvesting and also the selective flattening of crop circles. He transfers his gigantic designs from drawing board to land by using gridded drawings and simple surveying techniques.

This brings me to a few examples of "wonders of the ancient world." One that I have personally visited and that relates to the land art just mentioned are the huge designs on the plain of Nasca, Peru. The land there is hard-pan desert, with a scattering of small rocks on top, in appearance rather similar to the surface of the Moon. The grooves of the huge patterns laid out there are about eight inches wide and four inches deep. This is not very deep at all, and they would be long gone except that it rarely rains in Nasca. It rains so little that the locals leave a communal television sitting out on a bench, completely unprotected, in the city *zocalo*. (Of course, on the day I was there, it drizzled.)

Nonetheless, the plain probably averages less than an inch of rain a year, and the creation of the figures has been dated to a Nasca culture from the period 1 to 750 A.D. These figures include simple parallelograms as well as fish, spiders, monkeys, and other animals. It has been alleged that these can only be seen from the sky and were either made to signal visiting aliens or made by them, as the technique for making them is impossible to reconstruct.

Not true on any count.

The technique clearly was some straightforward form of a gridded scale drawing and manual surveying using string and poles to enlarge it. Hand tools would suffice to scrape out the shallow grooves. The patterns can be understood from the plain and easily viewed from standing on the surrounding hills. More reasonable interpretations see the lines serving as a calendar or as a constellation map though the possibilities have not yet been confirmed. I have seen a small ceramic pot in the National Museum in Lima that dates to the same place, time, and culture and is exactly the same shape as the fish design cut into the desert. (See my photograph opposite page 78, in *Are We Alone in the Cosmos?,* Bova and Preiss.) I would say that the lines must have a ritual purpose. In a recent article in *Archeology Magazine* (May/June 2000) Anthony Aveni shows fairly convincingly that these designs were tied up with the procurement of water, which was a big issue then, even though the climate was wetter than it is now. The lines might have mapped actual surface and subsurface sources, and in some cases might have been involved in the ritual of petitioning the gods for water.

So far I've discussed ancient two-dimensional art. What about three-dimensional wonders? The pyramids of Egypt and Peru, Stonehenge and the heads of Easter Island come to mind. Did the ancients have knowledge that has been lost to us? Are these works beyond the capabilities of the ancients, and therefore the work of aliens? Well, yes after a fashion, and no. We might not be able to specify the exact technique in every case, but we have investigated and found out the technique in some cases. In others several techniques feasible at the time of construction have been demonstrated to be capable of the required work. A specialized part of the field of archeology is now devoted specifically to the reconstruction of forgotten technology.

In the case of Easter Island, the actual hand tools used have been found in the quarries where the figures were cut, along with cut marks and partially completed statues. In Egypt and Peru, it has been demonstrated that the precise fitting of the gigantic stone blocks of the pyramids can be achieved by chalking the mating surfaces and rubbing the blocks together to assess contact. Then by further hand grinding and further contact tests, exact fits can be achieved. I have seen the results in Peru at Machu Pichu and other sites and the blocks are indeed huge and the fits precise. In both England and Egypt, it has been demonstrated (on TV yet!) that a group of 30 to 60 hardworking people could haul the stones of the pyramids and Stonehenge up dirt inclines. Then they can be positioned horizontally or dropped into a hole vertically, as the case may be. Of course, these processes are slow and would, especially in the case of the Egyptian pyramids, take many years of hazardous work by many people.

I hope archeologists will continue to explore and explain ancient civilizations. Let's not assume that aliens built them. Take (perhaps chauvinistic) pride in the great accomplishments of humanity. We deserve it!

8

I'M JUST A MATERIAL GUY

THE POSSIBLE DISCOVERY OF "A SUBSTANCE UNKNOWN TO SCIENCE"

What is now the single most famous UFO account dates back to the dawn of the modern UFO era, in 1947. It is the incident at Roswell, New Mexico. This event has spawned numerous journal articles, books, videos (notably the hoax called *Alien Autopsy*), and now two television series, *Roswell* and the famous *X-Files* (the latter of which I enjoy watching). The event began on a dark and stormy night, full of thunder and lightning, in July 1947. A rancher outside the town of Roswell saw a flash and heard what seemed like an explosion over a part of his land some distance from his house. The next morning the weather had cleared and he went out to take a look. He supposedly found a crashed saucer-shaped craft with a metallic surface, about 30 feet in diameter, in small pieces. He promptly informed authorities at nearby Roswell Army Air Force Base. The base sent an army crew that cordoned off the crash site. At a second, nearby site that was found a bit later, a saucer was found broken open with the bodies of several small aliens in crash seats and scattered on the ground, of which at least one was alive. Soldiers took away all debris and all the aliens, dead or alive. Meanwhile, a press officer at the Roswell air base issued a press release stating that there had been a crash and that parts of a disk or flying saucer had been recovered. The next day a news press release was issued by a press officer higher up the chain of command denying this account, and stating that the remains were those of a weather balloon. A reporter was allowed to interview a general and an air intelligence officer and take a few photos of what they claimed was part of the wreckage, which was laid out on the general's office floor.

The "real" material was supposedly taken back to the base, crated, then secretly airlifted to a base in Texas, and then secretly airlifted again to Wright-Patterson Air Force Base in Ohio. It is alleged to still be at Wright-Patterson, or alternatively at Area 51, secreted in an underground facility at that top-secret Nevada aircraft test range. An excellent recounting of the chronology and interpretation of the alien crash/conspiracy theory of Roswell can be found in *UFO Crash at Roswell* by Kevin Randle and Donald Schmitt.

It is also alleged that the United States has been using the recovered materials to reverse engineer advanced technologies and materials. However, just which advances are due to this source have not been identified in a definitive way. (Note that the following were recently listed in *The Day After*, by Col. Philip Corso, as quoted in an article by Robert Leach on page 51 of the June 2000 issue of *UFO Magazine*: night observation technology, fiber optics, computer chips, Kevlar and new metallurgical technology. Corso claimed to have had charge of the files that show these were some of the fruits of the Roswell crash. It should be further noted that these claims are unlikely to be true and unnecessary to assume. Fiber optics and especially computer chips have unbroken chains of human pedigrees, and their inventors, along with the woman who invented Kevlar, jealously guard their patents. There is no need or room for alien input in these histories.)

The Roswell scenario has been revisited again and again. It has been alleged that the whole incident was massively covered up by the U.S. government, starting from the Truman administration, through every political change, right through the present day, and that the cover-up has been managed by various committees, one of which was called MJ-12. Supposed documents written by MJ-12 have surfaced a number of times, especially at the start and end of the 1990s. The most recent appearance of supposed MJ-12 documents was found to be a hoax in the March/April 1999 issue of *UFO Magazine* (pp. 4-ff. and 70-ff.).

The reasons Roswell has become so famous revolve mostly around the fact that the army issued the first press release about saucers and then changed its story. This was only a month after the term "flying saucers" had been coined in the press to describe Kenneth Arnold's sighting. The press was hot for more, and because the army issued the story, it had great authority, at least for a short while. When the army retracted the release and gave its much more prosaic explanation, many refused to believe it. The sequence smacked of a cover-up. More recently, the military revised its story again and has now said that there *was* a cover-up, but it was of a test flight of a giant, high-altitude balloon intended to be used for Project Mogul, the previously mentioned top secret effort to monitor Soviet surface atomic tests. They were, of course, not willing to compromise this program before it got started. The conspiracy theorists are understandably not satisfied with this new explanation.

THE ROSWELL "CRASH SITE" REMNANTS AND OTHER ALIEN REMAINS

The majority of accounts of the Roswell crash state that the fragments recovered included several kinds of material with distinctive properties. In a 1988 interview by Randle and Schmitt of the rancher's son, he recalled finding a few leftover scraps at the crash site soon after the cleanup. These included thin sheet fragments that looked like metallic foil and were flexible to the touch but completely uncuttable, uncreasable, and jagged around the edges. If bent or creased, a piece would slowly unfold and return to its original shape (*UFO Crash at Roswell*, pp. 129, 131 and 132). In modern terms, the latter property is called shape memory. There is a small, known class of materials that have this property. One of these, a metal alloy called nitinol, has been used to make springs that coil and uncoil without hysteresis (progressive and permanent slight change in shape) and that permit the manufacture of very simple motors for mechanical work. The rancher's son also found pieces of balsa-like struts and of heavy-gauge monofilament fishing line-like wire.

It is also claimed that there were small metal struts, like I-beams in criss-cross section, that were broken into irregular lengths, about an inch high and up to a yard long. These, too, were said to be lightweight and flexible but not breakable or permanently bendable. Some of these pieces were said to have writing in an unknown alphabet. Published pictures of unknown authenticity show I-beams with lettering that looks more or less alphabetic (not pictographic) and is so far not attributed to any human system of written notation.

There is some disagreement about these material properties among those interviewed by Randle and Schmitt. The rancher's son gave the description of the resilient foil mentioned above. Those men who actually partook in the recovery said the sheet fragments were completely inflexible, and would not bend or dent even when jumped on or rolled over by a Dodge truck. Some pieces of the sheet had permanent curves, some were flat, and their edges were, by and large, clean and sharp. The fragments looked like metal and were said to feel like plastic (*UFO Crash at Roswell*, pp. 62 and 64).

The photos taken in the general's office at the "cover up" press conference clearly show a thin metal foil, but it looks like conventional aluminum foil, in places backed by paper. It is creased everywhere and ripped into large and small fragments with rough edges. There is string and what look like flat, thin wood struts incorporated along the edges of some of the foil fragments, all completely typical of weather balloons of the time. The fact that the fragments displayed to the press and photographed by the press appeared to be entirely conventional is explained by proponents of the alien crash thesis by saying that what the military showed off was exactly the remnants of a weather balloon, hauled in to buttress their cover story. (Incidentally, this also fits with the Air Force's admission that it was a cover story, but one intended to cover Project Mogul, which used balloons that may or may not have incorporated the existing ordinary technology.) The descriptions of the sheet fragments contradict one another in the area of flexibility and tearing properties. If there were both flexible and completely inflexible pieces, they must have come from different parts of the craft, or the descriptions must be regarded as unreliable.

There is another puzzling aspect to this account, one strictly technological. If the sheet material was flexible and it was part of the external shell or skin of the vehicle, that would pose a problem. Flexible skin on the outer sur-

Roswell, New Mexico, 1947. General Ramey (left) and Colonel Dubose display debris from a high-altitude weather device, insisting that the initial report of a crashed UFO was mistaken.

face of a flying machine would impose huge and variable stresses on the vehicle structure during high-speed flight and rapid acceleration in a planetary atmosphere like ours. Thus if the skin was flexible, the vehicle should not have been able to carry on operations inside our atmosphere.

While it is true that a flexible skin is used on U. S. submarines to reduce drag, it is a thin layer bonded to a conventional rigid steel shell. No evidence of such layering exists in any of the accounts of the crash recovery. It is also true that certain terrestrial materials are soft if they are stressed slowly but rigidify on sudden impact. Quicksand is one example. If you step on it slowly you sink in. If you slap it, it becomes rigid. Kevlar is relatively flexible to bending, but it is nearly rigid when bullets strike it. Whether this would work for an aircraft skin is questionable; the airflow, while it might be turbulent, would reach a nearly steady state during atmospheric entry and thus it is not easy to see how the flow itself would trigger the needed continuous rigidity.

The smooth cut/rough tear discrepancy is puzzling, too, unless two different materials were involved. The smooth cut might be explained if pieces of sheet were bonded together somehow, but no mention is made of remnants of edge adhesives or any riveting.

If the skin was flexible, another way out of the stress dilemma might be that the aliens possessed a technology to actively rigidify the skin. A tractor-beam technology might possibly be able to do this. But this would almost certainly require major space warp or black hole generator technology, and according to the crash accounts, there was no identifiable machinery of any kind in the wreckage. In fact, that is another puzzling aspect of the accounts. Where were the life support systems, power systems, and the propulsion equipment?

There is yet another problem posed by the crash recovery descriptions. If the vehicle had a metallic foil-like external skin, what prevented the interior from heating up to thousands of degrees during descent from orbit? A completely new metal alloy or metal-composite material would be needed to explain how the craft was able to survive in its assigned role to descend from orbit.

If there really was an alien ship crash at Roswell, it is past time to liberate some of the remains and lab reports under the Freedom of Information Act. The Cold War is over and alien invasion scenarios have been played out in the movies over and over to audiences of hundreds of millions. The human race is now so desensitized that almost any revelation that results would be an anticlimax.

TESTING THE FINDS FOR NON-HUMAN ORIGIN

If the material found at the crash site at Roswell still exists, it would be useful to subject it to various tests. These could include mechanical, electrical, magnetic, and heat property determinations. This would settle some of the questions that have been raised. It is, of course, possible that such examinations were made back in 1947 by the Army, but no documentation of this has ever been discovered. It would be useful to analyze the micro-structure of the materials, both in undisturbed portions and at the torn edges. Materials fail at the microscopic level first, and the microscopic failure pattern can give clues about how the materials were made in the first place. Such examinations would reveal whether the material is a sort of hybrid structure, such as fiberglass, or one of the carbon fiber composites. An electron-scattering experiment might reveal if the material has a crystalline structure, any long-range molecular or atomic order, or is amorphous like most glasses. The results could be compared to known materials and if unknown, this would be strong evidence for an extraterrestrial origin. Chemical analysis would yield the composition. If the amounts of different elements did not match any known pure materials or alloys, this would lend support to an extraterrestrial origin.

There are few other cases besides Roswell in which there are claims of significant physical alien material remains. The other category offered as evidence in a few cases is landing site material supposedly affected by the landing event. In a case mentioned in Peter Sturrock's book *The UFO Enigma*, soil samples were taken from a site near Trans-en-Provence in France, where a small ovoid is said to have landed briefly in January 1981. Lab tests done for project GEPAN/SEPRA (run by the French National Science Foundation) revealed that the soil was heavily compacted, that no organic traces as one might expect from chemical combustion were found, and that iron (without the usual amount of nickel or chromium as expected for steel) turned up. Traces of polymers, phosphate, and zinc were found. (The latter two are often found in fertilizer along with iron; the landing site was in a garden.)

In some cases, vegetation samples have been offered. In the case just mentioned, plant samples were obtained at increasing distances from the center of the discolored landing area. Analysis showed that there were significant drops in the amount of chlorophyll and beta-carotene, and that the drops were identical in samples taken both a day after and forty days after the landing event. The drops decreased in amount with distance from the center of the landing spot. Such decreases are associated with the normal aging process. In this case, there was no reason to expect normal aging was a factor. One of the chemists ran tests to try to find a quick aging process that could explain the observed change. He found that some but not all of the changes could be explained by exposure to a heavy flux of microwaves. He found no evidence of exposure to ionizing (nuclear) radiation (*The UFO Enigma*, pp. 98-99). GEPAN/SEPRA also investigated the 1982 "Amaranthe" case at Nancy, in France. A biologist reported that an ovoid hovered soundlessly about a yard off the ground in his garden for twenty minutes and then took off without heat or noise or smoke. Just before it left, the grass under it stood up straight. The investigators found that they could cause this effect in the lab by use of strong electric fields (tens of thousands of volts per meter). Perhaps this is relevant in discussing crop circles.

Chemical tests would presumably detect new elements if any existed in the samples. Radiation detectors will reveal whether radioactives were used or if there was exposure to them. In the case of the 1980 Bentwaters incident in England, the claim was made that the Geiger counters of the search team triggered as the craft took off. (At the same time generator-driven spotlights and two-way radios failed. This is curious because the Geiger counters, like the lights and radios, were electrically powered.)

Isotopic analysis might show whether the material came from another star system. Such an analysis requires heavy duty lab equipment and is not practical for field work. The theory behind this idea is that every star is made of matter that has a unique or nearly unique life history. It is a blend of elements made in the Big Bang origin of the universe along with elements transmuted by the fusion processes that make energy in the cores of stars and the bombardment processes that rapidly transmute elements up the mass scale of the periodic table when heavy stars go supernova. These materials mix into clouds in the turbulent interstellar medium and eventually the clouds collapse into stars and their planets. The process of planet condensation segregates elements at different distances from the parent star, giving a more or less progressive change in composition planet by planet, and as each planet forms, its elements tend to segregate internally too, the heaviest sinking to the center of the planet. This leads to complicated results, and in our solar system these still aren't fully understood. However, the isotope mixture from each alien world should be unique.

For example, uranium occurs in nature in mainly two isotopes, U-235 and U-238 (92 protons and 143 or 146 neutrons in the nucleus, respectively). The former is the more radioactive and the only form usable for bombs and reactors. These isotopes radioactively decay over time, respectively into lead-207 and lead-206. Isotopes of an element have notably different nuclear behavior from one another but almost identical chemical and other physical properties. This means ordinary chemical and physical processes involved in planetary formation and manufacturing are unlikely to affect the isotopic abundances. (This chemical identity is also why it is so difficult to perform the separation of U-238 from U-235, which is needed to make atomic bombs.) An analysis that showed an unusual ratio of isotopes of uranium or lead to isotopes of a number of the other heavy elements would suggest a different original mixture of elements, and thus a different star of origin.

NEW ELEMENTS ON AND OFF THE PERIODIC TABLE AND THEIR LIKELY PROPERTIES; ODD STATES OF ORDINARY MATTER: SUPER FLUIDS AND SUPERCONDUCTORS

It would be quite decisive if we were to find in purported UFO material an element we had never seen before. But what are the chances of that and what kinds of new elements might we expect? To answer these questions, first consider the known elements. They differ from one another in various chemical, physical and ultimately nuclear properties. The former two categories have been known and classified since ancient times, the last only since about 1900. In the late 1800s, a Russian chemist named Mendeleev designed a classification scheme for the then known elements that brought order to their chemical and physical properties.

His arrangement (one of many tried in that era) placed the elements in an incompletely-filled matrix, each row of which spread them out in order of increasing numbers of electrons (always equal to the number of protons in their nuclei), each column of which had similar chemical properties due to the number of outer electrons. He predicted that the elements in the remaining holes in the matrix would have properties that were the average of those of the surrounding elements. This prediction proved true, and his arrangement is the now-familiar Periodic Table of the Elements. As things stand now, every pigeon-hole in the table has been filled. Because each element differs by one electron (and proton) from its neighbors, there is no space left for new elements "in between" them. This means new additions must come at the light or heavy end of the table. But the light end is occupied by hydrogen, and because it has only one electron (and proton), nothing lighter is possible. So any new elements must be heavier than the known ones.

Nature probably creates these heavy elements in violent processes where subatomic particles aquire huge kinetic energies, typified by speeds more than a third the speed of light. The leading candidate for these creation events is the early moments of a stellar supernova explosion when neutrons at the edge of the stellar core are given gigantic energies and bombard the exploding outer two-thirds of the star as the inner third collapses to become a neutron star or black hole. Heavy elements from iron up through uranium are thought to be created this way. We have made a few heavier ones with proton counts up through 108 in microgram quantities by using conventional high-energy subatomic particle accelerators to bombard heavy atoms with protons. It was thought for many years that elements heavier than those could not be made due to accelerator limitations, but more importantly such super-heavy elements were thought to be unstable on theoretical grounds.

More recently, advances in theory suggested that a so-called island of nuclear stability might exist for elements with proton numbers above 114, perhaps up to the 120s. The expectation was that these would be unstable but might live more than a few seconds and thus become objects of laboratory study. Then advances in accelerator design allowed the use of whole heavy nuclei to replace protons as the high speed projectiles. The neutrons in such nuclei would form a short-range glue to overcome the long distance repulsion of the protons, which is due to their mutual electric forces. In 1999, two groups independently succeeded in synthesizing elements 116 and 118 using such accelerators, confirming some parts of the new theory. These new elements decay in milliseconds, and their physical properties are not well measured nor are their names settled. However, their creation suggests that perhaps there might be a way that nature does not provide to make much heavier elements. These might have extraordinary properties beyond the simple one of extraordinary density. Such super-heavy elements are likely to resemble their lighter cousins, and thus be metals or rare-earth-like, extremely dense and chemically reactive. These would not be attractive properties for spaceship structures. Possibly, they might be of use for atomic fuel or for making artificial black holes or wormholes.

There is another range of elements that is possible in theory, though only one has yet been made in the lab and none are known in nature. These are atoms that have muons in place of electrons, in orbits around normal nuclei of protons and neutrons. Muons have been dubbed heavy electrons because they have the same negative electric charge but weigh about 200 times more than electrons. Hydrogen-muon atoms have been made in the lab. Heavier ordinary atoms have temporarily housed muons mixed in with electrons. It is an open question whether heavier pure muonic elements would have the same periodic behavior as ordinary elements, but likely they would. It is however unlikely that they will be stable because muons themselves are unstable. It is also unclear how such matter would react if it comes in contact with ordinary matter. These muonic atoms would all have smaller radii than their normal relatives, and they would weigh a bit more (protons weigh 2,000 times more than electrons, but only 10 times more than muons.) One might guess that muonic matter might thus lead to more dense and more tightly bound atoms and molecules and thus stronger alloys. The strength increase might win out over the slight increase in density and make it suitable for flight applications.

All protons and neutrons are made of heavy particles with fractional electric charge called quarks. Recently, an atom smasher at Brookhaven Laboratory created for a brief moment a soup of quarks and gluons, the particles that hold quarks together and transmit the glue force, which is the source of the strong nuclear force. There are good theoretical reasons why quarks should only appear outside of protons and neutrons for brief instants. Normally, they are never naked in the universe. But it is barely conceivable that somehow a way might be found to make atoms out of them. The properties of such exotica are unknown.

Then there is antimatter. If enough anti-protons and anti-electrons could be created and made into stable atoms, a great new source of energy would be at hand. That's because when they contact ordinary matter, complete annihilation occurs, and there is 100 percent conversion to energy. But we can't yet create antimatter in useful quantities, and even if we could we have no safe way to contain it.

Aliens might have discovered not new elements, but new states of ordinary matter. We customarily deal with just three states: gases, liquids, and solids. In nature, we find some additional states. These include hot gases that are electrically charged, called plasmas. In these gases, free electrons roam and collide with the positively charged atoms that have lost them. Normally, plasmas only exist at temperatures of thousands or millions of degrees, as in the interstellar wind, inside stars, and the insides of controlled fusion experiments.

There is such a thing as degenerate matter. In nature, there are two main known versions. The first is found inside white dwarf stars. It is a mixture of a super-dense cloud of electrons that are packed in with a medium-heavy element like silicon. The density is in the range of thousands of tons per cubic inch. The other is neutronium—pure neutrons packed as densely as an atomic nucleus—the main constituent of neutron stars, which form during the collapse of some stars heavier than the Sun. The density is tens of thousands of times higher than white dwarf matter. Neither of these looks promising for a super-light spaceship structure.

More immediately useful are super-conducting and super-fluid materials. Super-fluid materials have the ability to flow without viscosity, that is, without internal friction. In principle they might be able to flow permanently without energy input after the starting push. Helium cooled to below 3 degrees above absolute zero has this property. In the lab it has been demonstrated that random fluctuations will cause a pool of liquid helium to spontaneously flow up the wall and over the rim of its vacuum flask container. The return of the (super-cool) blob, sort of. There have not been many applications of this due to the requirement to maintain such a low temperature. A room temperature super-fluid would probably have major applications in lubrication, in fluid flow, and perhaps in cooling. Nothing like this has been mentioned in UFO reports.

Recently and unexpectedly, a new class of complex ceramic materials was found to become super-conducting at temperatures much higher than absolute zero. Some now work at above liquid nitrogen temperatures. This is cold, but industrially simple, to produce. The first applications are for extremely sensitive magnetometers. High-temperature super-conducting wire will soon enter the market.

If a room temperature version can be made, then completely efficient electric transmission lines and super-conducting magnets will improve our power systems and allow high speed mag-lev trains. Room temperature super-conductors would be useful in spaceships, especially for controlled fusion rockets and for magnetic interstellar hydrogen scoops and debris deflectors, and possibly in creating warp drives. So far, no evidence of these has been found in UFO events.

SUPER-LIGHT/SUPER-STRONG MATERIALS AND THEIR BENEFITS: BOUNCING BUCKYBALLS AND BUCKYTUBES

Much more likely are improvements to the use of existing elements. Perhaps the most amusing earthly example is the discovery of a new form of carbon. Crystalline carbon is diamond, and amorphous carbon is found as graphite and carbon soot. The first new version is neither, but a sixty-atom molecule in the form of a spherical polyhedron with the shape of the seams of a soccer ball. It also looks like the structure invented in the 1930s by Buckminster Fuller called the geodesic dome. The chemist who first made this stuff cleverly named it buckminster fullerene, or buckyballs, for short. There are now known to be analogous spheres with fewer and more atoms, and also buckytubes with a structure that looks like that of the woven bamboo Chinese finger torture tubes. Buckytubes appear to have the possibility of making super-strong super-light fibers. But this is some time off. The buckyballs might turn out to be super-lubricants.

There is clearly great potential in the area of new compounds, alloys, and physical combinations. Extremely light, high strength materials have been devised using fiber/matrix composites and ceramics. Liquid crystals allow thin graphic displays. New techniques might allow computers to miniaturize down to molecular scale, well below microchip size. There have already been demonstrations of using DNA molecules to carry out computational problems.

NANO-MACHINES AS CONSTRUCTORS, STAR TRAVELERS, EMISSARIES, INFILTRATORS

If miniaturization could be applied to machines, some amazing and frightening possibilities might open up. K. Eric Drexler pioneered this area (see his *Engines of Creation*). He forecast the possibility of making machines that are on the scale of large polymer molecules. For comparison, DNA has a molecular weight in the billions. These might be lighter. Such nanomachines might be small enough to be released in your bloodstream to scour out your blood vessels, or be released at a trauma site to reconstruct organs. Or they might be made as assemblers to create microstructures on an industrial scale. As such they would assemble atoms into specific structures one by one. If first they are made as self-replicators, they could quickly create their own colony. The colony could create assemblers. The assemblers could create desired final machines or structures.

We have not yet reached that point. However, machines such as simple motors have been made using chip technologies that are smaller than the diameter of a human hair. It takes a powerful microscope to see the moving parts. Of course we do find analogs in nature. Rotifers collect food using tiny rotating wheels of sticky cilia. They are multi-celled invertebrates. You can see them in a drop of pond water using a microscope at 200x magnification. These living machines evolved to suit their niche.

It is natural to think of using nanomachines for tasks in manufacturing for space travel. A few nanobots weigh almost nothing and yet could be a high capability payload. For example, take a few of each type to space and turn them loose on a captured asteroid. They could convert the elements in the rock to useful equipment, starting with more of themselves, and then make parts for spaceships and colonies. A likely first goal would be extraction of oxygen from asteroid or lunar rocks. On an interstellar project they could assemble and maintain interstellar photon sails, which must be superlight and strong. Another role would be automatic mechanical failure detection and repairs on long voyages.

There are a number of potential problems to be overcome in this area. These include the means to provide nanobots with energy and also to provide them sensing and programming abilities. Most nightmarishly, there is the horror-movie possibility of a runaway nanobot infection, which could eat everything in sight including all matter dead and alive.

So far, there is no evidence that any UFOs have brought nanomachines with them for any purpose. However, nanobots represent an interesting possibility for an economizing interstellar civilization. Such a civilization might send nanobots in a very small ship. It would arrive and busily assemble macrostructures by extracting materials from surrounding soil and plants. Such a landing site would be an unusual scene and might morph into a blank gray patch, which would grow into a land rover and perhaps also a living ambassador. You would not want to stand close to the site until all the assembly tasks were completed. Otherwise, *you* might be liquefied and reassembled.

THE MACHINE, ELECTRONIC, AND BIO-VERSIONS OF "ALIEN VIRUSES"

A somewhat similar range of possibilities arises from consideration of alien viruses, both biological and computer. As travelers, they carry a lot of information at almost zero mass. The number of individuals can be in the millions for a single smear on a petri dish or a few digits of code. For comparison, note that our immune system, which consists of cells at small scale, recognizes more than a million different intruders based on our DNA codes. The care and feeding of viruses does require a host (viruses don't reproduce or "live" in isolation; they need a cell's machinery to do that). Still, the life support system becomes more modest, even if only reproductive cells or undifferentiated but fertilized embryo cells are on board. Upon planetfall, the incubators would go to work and grow new macroindividuals. This last step might, however, take a few years, if it is to result in a healthy and intelligent individual.

The computer viruses just need to reside in memory. Millions of viruses of different types could be written on a chip of a conventional sort. They would travel the universe and revive to do their work at planetfall. Perhaps computers and biology could be combined so that on arrival the alien child would acquire its memories from computer storage. Storing the contents (if not the capabilities) of the human brain should be feasible in the next twenty years. Incidentally, Fred Saberhagen's series of "Berserker" stories are all about machine intelligences (at macro scale) that are programmed to destroy all organic life. Fortunately, none of these have arrived here (yet).

9

ALIENS IN THE FAMILY

THE VARIETIES OF LIFE ON EARTH . . . AND ELSEWHERE

An earlier chapter described the range of alien body types reported by people who say they have visited with them or have been abducted. The range is surprisingly small. Is that what we would expect on scientific grounds? There are several ways to attack this question. Perhaps the most powerful way is to try to set down what anatomical functionality might be required for intelligent space-faring life to evolve. A counter argument of a sort can be investigated by considering the variety of life on Earth.

First, life needs a way to collect, analyze, and metabolize at least some of the most abundant energy supply in its environment. This is most likely electromagnetic energy radiated by its star but could also be thermal sources from the planet's interior. It could even possibly be energetic subatomic particles from radioactive decays, though no life form on Earth is known to consume the latter. It is easy to conceive of detectors for light; we call them eyes, and several forms have evolved on Earth. The two main ones are varieties of simple eyes with a focusing lens and single retina, like ours, and compound eyes with omatidia (eyelets) and multiple retinas, as found in lobsters, crabs, and insects. The analysis function is sometimes all in the brain and sometimes split between brain and eye.

The metabolization of the radiant energy varies. On Earth, sunlight is either directly absorbed by chlorophyll and converted to matter in plants, or indirectly transformed as those plants are eaten by herbivores and converted to animal flesh. In some cases, this flesh is eaten by carnivores. Ocean vent life takes in energy without sunlight through the minerals and heat released from the vents that are open below to upflowing lava from the upper mantle. The range of vent life is startling, but at the bottom of the food chain are single-celled bacteria-like *archaia* that live off sulfur instead of oxygen. These are thought to be among the first of all living things to have evolved on Earth. But time has allowed a large array of specialized digestive schemes to evolve among both single-celled and multi-celled organisms, and blood composition varies. For example, though blood carries oxygen and removes carbon dioxide in many complex organisms, blood composition varies. Iron plays an important role in the blood of many animals (including ours, of course), but in some invertebrates copper plays that role in a molecule called hemocyanin, which is blue when oxygenated, rather than red, as ours is.

The ability to grasp and manipulate things in the environment is probably a necessity for creating technology, and thus likely a requirement for any star-faring species. We have, of course, a pair of mobile grasping limbs terminating in small jointed digits with an opposable digit and slow growing, easily trimmed nails. These allow small things to be picked up and manipulated. For us to be able to properly grip without dropping or crushing what we hold, we need, and have, a pressure-feedback system—our sense of touch. If you think that is not essential, try picking something up when your hand is "asleep" and your sense of touch is deadened. But don't pick up anything fragile.

Opposable digits are not the only workable grip system. Other examples include the trunks of elephants and the tentacle-sucker system of octopuses. In fact, the octopuses show excellent ability to grip, orient, and unscrew things, even though they have no joints for leverage and no rigid skeleton. They get along on muscles and suction. Their tentacle tips, the smallest parts, are not important for this, and thus they cannot manipulate things as tiny as we can.

For any manipulation, visual feedback is important. Almost any animal able to manipulate objects can watch what it is doing with binocular vision and because of its body arrangement.

Another thing important to the ability to manipulate objects is the freedom of the organs of manipulation from other tasks most of the time. Thus, being upright (humans and elephants) or buoyant in a supporting fluid with other means of propulsion (octopuses) is likely to go hand in hand with the ability to develop technology.

One could argue that this logic is unnecessarily biased in favor of the development of *inorganic* technology. There might be a way to craft starships entirely out of living things, or even that life might have evolved to intelligence in interstellar space, able to survive the vacuum and eons-long transits while in hibernation. In an excellent novel by James Blish, *A Case of Conscience*, a tree on a distant planet evolves a steam-driven seed-ejection system that is powerful enough to allow the seed to escape its solar system, and the seed is hardened enough to survive to next planetfall. Larry Niven and Jerry Pournelle, in a number of short stories and novels, have postulated a species called the Outsiders, which perpetually travels between the stars on spars open to the vacuum, living and thinking at cryogenic temperatures, carrying on a trade in intellectual property. Such possibilities are not ruled out, but given the relative benignity of a range of planetary environments, I would argue that complex life is more likely to evolve on a planetary surface than anywhere else, and that inorganic technology is the most likely route life can take to an organized society with the resources necessary to develop interstellar travel. And for this, the ability to manipulate is needed.

We are notably bilaterally symmetric—left eye, right eye; left leg, right leg, etc. This aids our ability to manipulate on both large and small scales. Bilateralism has evolved throughout the animal kingdom and can be considered a generally successful adaptation. But it is not necessary. Again consider the octopus. It is radially symmetric with respect to limbs. However, its tentacles may be and usually are used opposed to one another for manipulation. Niven and Pournelle, in *The Mote in God's Eye*, imagine a species that has evolved with three arms, of which two are on one side and one on the other. This same species has engineered specialized sub-species for various purposes, including miniatures that serve as engineers and construction crews with extremely fine ability to manipulate.

Because we live in a sound conducting medium, but sound is intermittent and a low density energy source, sound is not our energy supply but primarily carries information about changes of state in the environment. These might be from inorganic causes, such as earthquakes and thunder, or organic in origin, such as howls and whispers. Life here evolved to detect and make sounds as a result of their natural availability and survival value. It seems likely this would happen on any planet. Our senses of smell and taste are closely related to each other and, like sound, do not supply us with metabolic energy, but rather with signals about internal and external changes. Both senses are based on chemical gradients in time and space, and because molecules can disperse in any fluid, liquid or gas, they are mobile everywhere in a planetary environment that is not totally solid. I would expect any alien to be able to taste and smell, probably for all the earthly uses, but the detailed spectra of chemical sensitivities might be quite different.

Might there be additional senses, and if so, what would they be? The ability to detect magnetic fields might develop. Any solid planet with the heavy elements needed for life as we know it is likely to have an iron core and a global magnetic field. Some terrestrial bacteria have this ability. It allows them to orient to and travel along the Earth's magnetic field lines. Some birds apparently have this ability and use it to navigate intercontinentally. Some insects see ultraviolet light. Dolphins and bats use sound echo location, natural sonar, to locate things around them, and dolphins, at least, can see inside solid objects using sound. Some animals have much more acute smell than we do, sharks and dogs among them. This provides location information but not shape. Perhaps alien animals might evolve a sense of pressure sufficient for an organism to accurately sense its altitude in atmosphere or its depth in ocean. Or maybe aliens might evolve a highly accurate sense of time that would allow precision internal timekeeping.

Many animals combine sensory information of several types to do complicated tasks. For example, vision and touch and our sense of internal fatigue allow us to estimate whether we can catch fleeing prey, or whether we will beat a car to an intersection. Aliens might evolve unusual combinations of senses. Such combinations would not require new senses, only different wiring in the brain.

WILL WE SHARE BIOCHEMISTRY? THE SAME GENETIC MOLECULES? CARBON VS. SILICON?

Chemistry is fundamental to all organic life. It is at the heart of metabolism, heredity, body structure, behavior, reproduction, and all the senses. Life has evolved on Earth in niches that run from deep ocean ridges in total darkness at hundreds of atmospheres and temperatures above that which boils water, to bone-dry Antarctica at temperatures way below freezing, to the still little known biota of the near-vacuum of the stratosphere. Recently, discoveries have also shown that there might be a micro-biota living in the hot rock below the ocean floor, perhaps also under the continents, which might contain more biomass than all other life forms put together. Biochemistry, contrary to the mid-twentieth-century understanding, is not so terribly fragile and can thrive in these extreme conditions, as well as the normal ones we enjoy. The new field of investigating extremophylic life is exploding now, precisely because it is seen as yielding insights to our search for life on other planets and the origin of life on ours.

All of this life is founded on carbon-based molecules. These include the hereditary code for all life, the long chains of DNA and RNA, which are carbon plus hydrogen, nitrogen, and heavier elements like phosphorus and sulfur. These, in turn, control all cell processes that take short-chain amino acids and build them into longer proteins, becoming flesh and consuming energy. The code is universal here. The amino acids are, too.

I think there is a strong argument that alien life will also be based on DNA-type molecules and amino acids similar to or identical to ours. This does not necessarily imply that the aliens will look like us or anything on Earth. The reason they might look very different is the immense number of possibilities inherent in the genetic code of DNA, which in complex animals has an information capacity of billions of bits, equal to the contents of the *Encyclopedia Britannica*. Just consider all the different species that have evolved here over geologic time. There is nearly infinite room for other variations in the immense information capacity and mutational possibilities in DNA, or any similarly sized master molecule.

The idea has floated around that another atom might substitute for carbon as the basis for life molecules. The element most often mentioned is silicon (the semiconductor, as in Silicon Valley), because it has nearly the same binding possibilities at about the same energy levels as carbon. I think this is unlikely because silicon is much more abundant than carbon on Earth, yet over the eons silicon-based life has failed to evolve here. If it were feasible to happen, there should be silicon life here. Silicon is incorporated in the structures of some life forms (sponges and diatoms, for example), but it is not the basis for it.

There is another reason, a more direct one, to think that carbon-based molecules, including amino acids like ours, are universal. That is the fact that studies of a number of carbonaceous chondrite-type meteorites have yielded amino acids identical to ours. These meteorites have ages of 3.5 to 4.5 billion years. They are leftovers from the early solar system and almost certainly predate its time of formation. Their base materials certainly do, having condensed out of a pre-existing interstellar cloud. That these amino acids are truly from space and not earthly contaminants is well established by the fact that they occur almost equally in left-handed and right-handed versions, while on Earth left-handed amino acids dominate. Furthermore, recent observations of light from interstellar clouds have shown that it is predominantly polarized in the left-handed sense, which buttresses the idea that the left-handedness dominates the building blocks of life from before solar system formation. These results strongly suggest both that carbon-based life must be widespread, and that our amino acids are universal, and, by implication, DNA, organized the same as ours with the same base pairs as ours (or it would not handle the same amino acids), is also universal.

*[NEWSFLASH (Dateline-San Francisco, 6/17/2000): A report published today states that radio astronomers have just announced the discovery in space of glucoaldehyde, a simple sugar of eight atoms consisting of carbon, oxygen, and hydrogen. This small sugar can combine with other atoms to form the more complex and biologically important sugars ribose and glucose. Ribose, please note, is a basic building block of RNA (*ribo*nucleic acid). The glucoaldehyde was found in an interstellar cloud halfway to the center of*

our galaxy. This is even stronger support for the idea that our type of RNA and DNA is universal.]

Some might object that our food might be poison to aliens, and vice versa. This is certainly a possibility, given that many things on Earth are poisonous to us. Assuming that aliens run on amino acids like ours and evolved on a rocky planet with a reasonably similar range of heavy elements, they might nonetheless survive on food they find here because of what appears to be a universal principle of life chemistry: organisms are homeostatic in most respects. That means they tend to maintain their system biochemistry in a steady state, no matter what they take in as food or what disturbing influences they suffer. Even though the chemical composition of each type of food, and in general each specimen of a food varies from every other, every organism in effect takes what it needs and excretes the rest. Deficiencies are sometimes harder to deal with, but most animals can smell their way to the foods that fill in those deficiencies. Animals can find salt, for example, by smell.

Of course, all of this is not proof. It is possible to conceive of other elements participating in the base pairs of DNA, or of more than four bases in the code. Again, however, the possibilities for such differences exist here because there is a full range of elements, and yet they did not happen. Or if they did, the organisms they constituted all went extinct long ago.

COULD WE BE CROSS-BRED WITH ALIENS, AS MANY ABDUCTEES REPORT WAS TRIED ON THEM?

I am now going to take a position that may surprise some readers. Because I believe that amino acids identical to ours are widespread in the galaxy, and because it is likely that only our type of DNA can manipulate them, I think aliens and alien life in general will turn out to be biochemically like us. Therefore those abductees who claim that aliens are running a cross-breeding program that mixes us and them cannot be proved wrong *in principle*. In principle, I think such a thing would be possible.

It used to be thought that interspecies exchange of genetic material was impossible. But evidence in nature has been found that bacteria have transported some genes and gene fragments from one species to another. Our

DNA contains code that is also found in some viruses. The latter is evidence that viruses were our ancestors or that, more likely, they infected our ancestors. Our cells contain structures called mitochondria—their energy factories—that billions of years ago were almost certainly independent bacteria that took up a symbiotic relationship with other, larger cells, and stayed on. Mitochondria contain DNA independent of nuclear DNA, which descends only from the mother. So our body chemistry has "alien" bits in it of a sort.

Would such a cross-breeding program be simple? No. It would be immensely difficult. Recently, we have genetically engineered the genes for the human immune system into mice. This cutting edge technique allows us to test immune related drugs in mice and expect to get a response reliably predictive of the human response. This, however, is mere child's play compared to a human/alien crossbreed. The latter would be equivalent in difficulty to trying to cross humans with terrestrial species outside our immediate families. It would be at least as difficult as crossing humans with seals or dogs, and more likely as difficult as crossing us with reptiles, fish, or birds—or even with some of the complex creatures that don't fit in any known phylum that went extinct shortly after appearing on the terrestrial scene 550 million years ago (see Stephen J. Gould's *Wonderful Life*, or Simon Conway Morris's *The Crucible of Creation,* for their stories).

First, no ordinary cross-breeding by mating or grafting would work. Normal fertilization would fail; the graft would not take. Embryo jumbling would almost certainly fail due to the genetic distance between species. The one case we have where embryo-bashing succeeded was that of the "geep," carried out at the University of California at Davis in the 1987. The embryonic cells of a sheep and goat were combined and implanted in a ewe, where they grew into a fetus, which was born and raised to adulthood. The geep was female, with the head and neck of a goat and the body of a sheep, and was later able to carry an implanted sheep embryo to term. But it was sterile. And this example represents a combination of *very* closely related species.

The only techniques with a chance of success are the new techniques of genetic engineering. Here the knowledge we are missing includes, but is not limited to:

1) How changes to DNA affect the form of the species. Recent discoveries about the HOX gene have begun to reveal the secrets of developmental changes and the placement and numbers of limbs and organs.
2) How DNA controls metabolism, in detail.
3) How DNA structure affects the biochemistry of reproduction.
4) How the DNA controls the structure and operations of the immune system.

This last item is a matter of intense interest to us already as it is key to understanding cell and organ transplant rejection. The normal human immune system recognizes from the fetal stage more than a million different substances that it can defend against. Until the late 1800s so little was known about this that people died when transfused with the wrong blood type. The very existence of immunologically different blood types was not then known. Transplants from other humans, with only a few exceptions, and cross-species transplants, are eventually rejected with a usually fatal reaction. We give transplant patients immune suppressing drugs to prevent that, but all these drugs knock down the whole system, leaving the patient open to just about any infection.

The distance from our current state of knowledge in this area to the state that would permit interspecies cross breeding is *immense*. But I don't know of any physical or biological law that such a manipulation would violate. And any aliens have likely been around a lot longer than we have.

Having taken the leap that it is in principle possible, do I think that such a program is now underway? I don't think so, but for reasons unrelated to the problems of genetics itself. Even if we are currently being visited by aliens, the logic of their carrying on such a program—with the apparent uncertainty and the crude surgical techniques reported by abductees—completely escapes me. The objection that "aliens are aliens" in thought and deed and thus unknowable is not viable. Surely any starfaring species has command of all the genetic knowledge it needs to do exactly what we have described above, and will have done any cross-breeding either with species native to their home planet or will have the knowledge to construct new species (like us) from scratch by molecular assembly of DNA or their analog of it. Their techniques would likely never involve any substantial sampling by penetration, much less failures to correctly anesthetize the subject during such a procedure.

COULD WE BE DESCENDANTS OF ALIEN SEED SENT TO EARTH? OF INTERSTELLAR GARBAGE DUMPING?

Science fiction and some abductee accounts suggest that we are really descendants of alien species. In some cases our ancestors allegedly landed here 10,000 to 100,000 years ago. In others, we are said to have evolved from alien trash. Neither of these scenarios is likely, based on examination of the evidence, both genetic and geological.

The genetic evidence was mentioned above: by sampling contemporary species we can trace genetic consanguinity and easily show that all terrestrial life shares some parts of its DNA, and we can estimate how long ago each species had an ancestor in common with each other species. Then, with radiological absolute dating techniques, we can show that the ancestors of living species seem to fit this picture at least in some cases. In the case of amber-preserved insects we can get both a geologic age and, from extracted DNA (*à la* Crichtons's *Jurassic Park*), a molecular DNA age.

While it is true that the fossil record of early humans is relatively poorly known, it is well enough sampled to show a reasonably clear line of descent back into the pre-primates of about 6 million years ago. And it is reasonable to expect that the picture will eventually become more or less complete. The million-plus-year-old footprints of adult and child preserved together in volcanic ash in Africa clearly demonstrate that fundamental human family values go way back. There is just no room for alien interlopers.

Given the evidence of 2.8-billion-year-old strands of algae cells from the Gunflint formation and the stromatolite fossils (compacted algal colonies) found around the world that date back to 3.8 billion years ago or older, which both look essentially identical to contemporary examples, it is hard to believe that life began other than as the result of the rain of cosmic amino acids and other chemicals onto Earth after it cooled down following its formation 4.5 billion years ago. It is the simplest explanation.

PAN-SPERMIA, COMET-FALLS, MANUFACTURED INFECTIONS, AND OTHER NOTIONS OF INTERSTELLAR EPIDEMICS

At the turn of the twentieth century the Swedish chemist Svante Arrhenius put forth the theory of pan-spermia. It is roughly similar to the idea of the cosmic catapult tree from Blish's *A Case of Conscience*. Arrhenius suggested that life might evolve forms so hardy that they could survive the voyage between the stars, and that over time life would spread from planet to planet and star system to star system. This now seems relatively unlikely though not absolutely ruled out. The detection of amino acids in meteorites makes it likely that what spreads are the building blocks of life.

The conditions of space seem too hostile to allow true life to evolve. Each level of complexity involves more interactions for success, and the low densities of matter even in dense interstellar clouds, along with the deep chill there, make such interactions improbable.

This did not stop Sir Fred Hoyle and his colleagues from suggesting back in the 1970s that some of the viruses and bacteria that infect our world have evolved in outer space and that they have rained down on us in cometary debris as it falls to Earth. To some extent such space junk falls all the time. But the notion that bacteria and viruses can evolve on comets seems unlikely on the grounds outlined above, though it should be noted that comets are composed partly of water ice. But that ice probably sublimes directly to gas as the comet is heated up as it approaches the Sun, bypassing the liquid state.

Could alien incoming viruses and bacteria infect us? Since I have argued that they would likely be made of the same amino acids and similar DNA, it is not ruled out. But the probability of infecting us, having lived for eons in a frozen state, seems unlikely. Think of all the viruses and bacteria already here that do *not* infect us.

THE NATURE OF THE ALIEN PSYCHE

To have survived a long time aliens almost certainly must have found principles of ethics that seek to minimize suffering. Sure it's conceivable that sadomasochistic aliens might exist. But since pain signals damage to an organism, maximizing pain for pleasure is not a survival strategy likely to win out in the end. A species will die if it kills off too many of its own. Individuals will surely try to protect their young and let them grow to reproductive age. Thus aliens can reasonably be expected to have found a path or paths to peace among themselves, if not toward others. Their medicine should be founded, like ours, on the principle of "do no harm."

All of this is a relatively weak constraint on probable alien behavior. If we think about all the animals on Earth, we can see a range of strategies around food, predation, defense, and parenting that is quite wide in its details. In fact, there is enough of a range of niches and behaviors here that I believe when we finally meet aliens their behavior will be understandable to us and will fall within a form we have already observed in some species on Earth. That still leaves a pretty wide field. Consider our slaughter of "lesser" animals such as buffalo and whales. Or even the extreme intransigence of the North Koreans toward the South Koreans and the U.S. since the cessation of hostilities fifty years ago. We might find aliens difficult to deal with.

If, on the other hand, they have adopted life-affirming principles, it seems likely we should be able to deal with each other on a fair basis, no matter how far ahead of us they are technologically. Then they should enjoy equal rights with humans, and should be treated as full citizens of our world while they visit us, or live among us. On that day I would also hope and expect that they would treat us with respect and offer us technical and philosophical partnership as appropriate. I hope with all my heart that I will live to see that day.

References, Recommended Readings, and Useful Resources

PRINT MEDIA

Abbott, E.A. *Flatland.* New York: Dover Thrift, 1992. (first published 1884).The great classic of dimensional visualization.

Adamski, G. *Flying Saucers Have Landed.* New York: Abelard-Schulman, 1953.

Andrus, W. H. and Hall, R.H. *MUFON 1987 International UFO Symposium Proceedings*. MUFON, Seguin, TX, 1987.

Aveni. "Solving the Mystery of the Nasca Lines." *Archeology Magazine* (May/June 2000) : 50ff.

Barr, S. *Experiments in Topology*. New York: Dover, 1989. Hands-on experiments you can do at home. Lots of thought problems.

Becker, W. *Link.* Avon, New York, 1998, The bibliography on pp. 413ff. The bibliography gets you into arcane alien precursor theories.

Bester, Alfred. *The Stars My Destination*, New York: Bantam, 1970.

Blumrich, J.F. *The Spaceships of Ezekiel.* New York: Bantam, 1974. Incredibly detailed analysis of Ezekiel's wheel as alien spaceship.

Bova, B. and Preiss, B. *Are We Alone in the Cosmos?* New York: Simon and Schuster, 1999. The scientific search for extraterrestrial intelligence. I am the "science editor" on this book.

Burke, J. G. *Cosmic Debris*. Berkeley: University of California Press, 1986.

Bryan, C.D.B. *Close Encounters of the Fourth Kind.* New York: Penguin, 1996. Balanced and detailed account of abduction, including detailed interviews with abductees.

Cavelos, J. *The Science of Star Wars*. New York: St. Martin's Press, 1999.

Cohen, S., et al. "How to Fake UFO Photos." *The Skeptic*, vol. 6, no. 4 (1998) :102ff. Simple, clever, effective.

Cole, K.C. "Escape from 3-D." *Discover* (July 1993) : 52ff.

Drexler, K. E. *The Engines of Creation*. New York: Anchor-Doubleday, 1987. The seminal work on nano-machines.

Gould, S. J. *Wonderful Life*. New York: W.W. Norton, 1989. Beautifully written and illustrated discussion of the new information on the explosion of multicellular life.

Green, B. *The Elegant Universe*. New York: Vintage-Random House 2000. A carefully crafted explanation of the latest views on the multidimensional universe, string theory and their connections to relativity and quantum theory.

Greenler, R. *Rainbows, Halos and Glories*. Cambridge: Cambridge University Press, 1980

Hamilton, B. "Phoenix Lights: Part II". *UFO Magazine* (June 2000) : 22ff.

Herd, S. *Crop Art*. New York: Harry Abrams, 1994. Spectacular work, beautifully photographed.

Hertz, J. H., ed. *The Book of Ezekiel*, The Pentateuch and Haftorahs. London: Soncino Press, 1989.

Hogan, J.P. *Cradle of Saturn*. Riverdale: Baen, 1999. The bibliography pp. 525ff. Another clever book from a master of hard sci-fi. Here he reveals himself as a Velikovskian catastrophist. Overboard, but a good bibliography for those who wish to explore this mistaken theory further.

Kaku, M. *Hyperspace*. New York: Anchor-Doubleday, 1995.

Kraus, L. *The Physics of Star Trek*. New York: Harper-Collins, 1996. Straightforward, clear and entertaining.

Luminet, J-P. et al. "Is Space Infinite?" *Scientific American* (April 1999) : 68ff.

Mack, J.E. *Abduction*. New York: Ballantine, 1995. Exhaustive introduction to abduction case studies, but it goes almost nowhere.

Mallove, E. and Matloff, G. *The Starflight Handbook*. New York: John Wiley, 1989.

Minnaert, M. *Light and Color in the Open Air*. New York: Dover, 1954. A classic in its field.

Morris, S.C. *The Crucible of Creation*. Oxford: Oxford University Press, 1999. An excellent survey and partial rejoinder to S. J. Gould's *Wonderful Life*.

Naud, Y. *UFOs and Extraterrestrials in History, Vols. 1-4*. Geneva: Editions Ferni, 1978. A huge and extraordinary survey of UFOs and abductions from ancient times to nearly the present. Not very selective and rather uncritical, but still a great resource.

Pickover, C. A. *Surfing Through Hyperspace*. New York: Oxford University Press, 1999. Great illustrations of shapes in hyperspace.

Pringle, L. *Crop Circles*. New York: Harper-Collins, 1999. Wonderful photos of great art, but uncritical alien-centered views.

Randle, K. D. and Schmitt, D. R. *UFO Crash at Roswell*. New York: Avon, 1991. Seems to be careful and complete.

Randles, J. and Warrington, P. *Science and the UFOs*. New York: Basil Blackwell, 1987. Considers interesting UFO and abduction cases mostly from a psychological point of view.

Robin, T. *Fourfield*. Bullfinch. 1992.

Rucker, R. *The Fourth Dimension*. New York: Houghton Mifflin, 1984. The most carefully reasoned, clear, and well illustrated book on the subject.

Sagan, C. and Page, T. *UFOs—A Scientific Debate*. New York: Barnes and Noble, 1996. Reprinted from 1972; essays by scientists on various cases and subjects.

Shaw, Bob. *Night Walk*. New York: Avon, 1967.

Steiger, B., ed. *Project Blue Book*. New York: Ballantine, 1976. A clear discussion of the U.S. government UFO studies up to its date.

Sturrock, P. *The UFO Enigma*. New York: Warner Books, 1999. The most up-to-date and critical evaluation of the U.S. government UFO studies along with carefully considered suggestions for future scientific studies.

Thorne, K. *Black Holes and Time Warps*. New York: W.W. Norton, 1995. A classic of exposition of this great subject, with excellent examples and illustrations.

Wilson, J. "Six Unexplainable Encounters." *Popular Mechanics* (July 1998): 62ff.

Zeilinger, A. "Quantum Teleportation." *Scientific American* (April 2000): 50ff.

VIDEOS

Fox. "Best Evidence for UFOs: Part 2"

WGBH/NOVA. "Kidnapped by UFOs?" WGBH, Boston, 1996

Fox. "Alien Autopsy: Fact or Fiction," 1995

WEB SITES

http://Space.com/area51/index.html

http://www.scientificexploration.org (the site of Sturrock's Society for Scientific Exploration)

http://www.ufocenter.com (the NUFORC site)

http://www.cufos.com (the J. Allen Hynek Center for UFO Studies)

http://www.fufor.org (the Fund for UFO Research)

http://www.seti-inst.edu (the SETI Institute)

http://foia.cia.gov (the CIA Electronic Document Release Center)

http://www.foia.fbi.gov (the FBI Freedom of Information Act Document Electronic Reading Room)

http://www.ufomag.com (devoted to UFOs and the paranormal)

INDEX

INDEX

INDEX

INDEX

INDEX

INDEX

INDEX

INDEX

INDEX

INDEX

INDEX

About the Author

William R. Alschuler has a Ph.D. in astronomy from the University of California at Santa Cruz, and extensive college teaching experience in the sciences, holography, Lippmann photography, energy conservation and solar building design. He is founder and principal of Future Museums, a firm that consults on the design of exhibits and museums with a science or technology content. He has recently served as consultant to the California Science Center and the Getty Education Institute for curriculum development that combines art and science. He has authored or edited for Byron Preiss Publications *The Microverse, UFOs and Aliens, First Contact, The Ultimate Dinosaur*, and, most recently, *Are We Alone in the Cosmos?* He is currently a science professor at California Institute of the Arts and lives with his family in San Francisco.